Environmental Science

Environmental Science

THIRD EDITION

Active Learning Laboratories
and Applied Problem Sets

Travis P. Wagner, Ph.D. and Robert M. Sanford, Ph.D.

Department of Environmental Science & Policy
University of Southern Maine

VP AND EDITORIAL DIRECTOR	Petra Recter
SENIOR EDITOR	Alan Halfren
EDITOR-SCIENCE	Jennifer Yee
EDITORIAL MANAGER	Judy Howarth
CONTENT MANAGEMENT DIRECTOR	Lisa Wojcik
CONTENT MANAGER	Nichole Urban
SENIOR CONTENT SPECIALIST	Nicole Repasky
PRODUCTION EDITOR	Indirakumari Siva
PHOTO RESEARCHER	Billy Ray
COVER PHOTO CREDIT	© Courtesy of Robert M. Sanford

This book was set in 10/12 Times LT Std Roman by SPi Global and printed and bound by Quad/Graphics.

Founded in 1807, John Wiley & Sons, Inc. has been a valued source of knowledge and understanding for more than 200 years, helping people around the world meet their needs and fulfill their aspirations. Our company is built on a foundation of principles that include responsibility to the communities we serve and where we live and work. In 2008, we launched a Corporate Citizenship Initiative, a global effort to address the environmental, social, economic, and ethical challenges we face in our business. Among the issues we are addressing are carbon impact, paper specifications and procurement, ethical conduct within our business and among our vendors, and community and charitable support. For more information, please visit our website: www.wiley.com/go/citizenship.

Evaluation copies are provided to qualified academics and professionals for review purposes only, for use in their courses during the next academic year. These copies are licensed and may not be sold or transferred to a third party. Upon completion of the review period, please return the evaluation copy to Wiley. Return instructions and a free of charge return shipping label are available at: www.wiley.com/go/returnlabel. If you have chosen to adopt this textbook for use in your course, please accept this book as your complimentary desk copy. Outside of the United States, please contact your local sales representative.

ISBN: 978-1-119-46259-0 (PBK)
ISBN: 978-1-119-46258-3 (EVALC)

Library of Congress Cataloging in Publication Data:

Names: Wagner, Travis, author. | Sanford, Robert M., author.
Title: Environmental science : active learning laboratories and applied
  problem sets / by Travis P. Wagner, Ph.D. and Robert M. Sanford, Ph.D.,
  Department of Environmental Science & Policy University of Southern Maine.
Description: Third edition. | Hoboken, NJ : Wiley, [2018] | Includes
  bibliographical references and index. |
Identifiers: LCCN 2018009378 (print) | LCCN 2018013588 (ebook) | ISBN
  9781119462576 (pdf) | ISBN 9781119462552 (epub) | ISBN 9781119462590 (pbk.)
Subjects: LCSH: Environmental sciences—Study and teaching. | Environmental
  sciences—Problems, exercises, etc.
Classification: LCC GE70 (ebook) | LCC GE70 .W34 2018 (print) | DDC
  628.078—dc23
LC record available at https://lccn.loc.gov/2018009378

The inside back cover will contain printing identification and country of origin if omitted from this page. In addition, if the ISBN on the back cover differs from the ISBN on this page, the one on the back cover is correct.

"Those who dwell, as scientists or laymen, among the beauties and mysteries of the earth are never alone or weary of life."

from *The Sense of Wonder* by *Rachel Carson*

(1907–1964)

Acknowledgments

Special thanks are due to our University of Southern Maine colleagues, Prof. Sarah Darhower, Dr. Lisa Hibl, Dr. Daniel Martinez, Dr. James Masi, Dr. Joe Staples, Dr. Karen Wilson, and Prof. Linda Woodard. Thanks to Lee Ann McLaughlin. Special thanks to our former Teaching Assistants Victoria (Tori) P. Hill and Shereen Toolabi. Thank you to Stephanie Ayotte for letting us use her laboratory reports. Thank you to Kelsey Earley. Thanks to Dr. Samantha Langley of Northern Kentucky University. We appreciate the patience of our students as we tried out various versions of these laboratory and homework activities. Helpful comments came from the external reviewers provided by John Wiley & Sons. Thanks to David Sanford, P.E., and USM library staff Max Calderwood and Ed Moore. Thanks to SENCER (Science Education for New Civic Engagements and Responsibilities) for encouragement of civically minded science education.

Preface

Environmental Science: Active Learning Laboratories and Applied Problem Sets is designed to introduce learners to the broad, interdisciplinary field of environmental science by presenting specific labs that use natural and social science concepts to varying degrees and by encouraging a hands-on approach to understanding the impacts from the environmental/human interface. The laboratory and homework activities are designed to be low cost and to reflect a sustainability approach in practice and in theory.

In addition to the overall approach and design, this text:

- Uses innovative learning techniques, such as problem-based, active, and critical learning. Group and cohort paths to knowledge are encouraged. As part of this approach, we stress learner initiated inquiry and experimentation, as well as an emphasis on civic responsibility in environmental science.

- Develops a variety of topics that mirror the variety of subjects found in environmental science, including urban ecology, global impacts, air pollution, solid waste, energy consumption, soils identification, water quality assessment, and the scientific method.

- Encourages learners to grasp the big picture by relating the lab activity to real-life conditions and their individual contributions to environmental problems. We have individual measures and descriptions, but we also nurture application of this learning to the larger ecological picture.

- Develops a variety of techniques, including traditional laboratory activities, field exercises, Internet research, calculations/extrapolations, and critical analysis. Because the pursuit of real-world environmental science involves all of these components, so do our lab activities.

- Emphasizes the improvement of written and other forms of communication. So much of science has become participatory, particularly in making decisions about its application (i.e., environmental policy). We provide ways for the learner to discover that the communication of scientific information is as important as the acquisition of scientific knowledge.

- Contains relevant problem sets that can be used as labs, lab supplements, or homework assignments for environmental science lectures.

Contents

Preface   vii

Introduction   xii

PART 1   General Information   1

    Laboratory Health and Safety Procedures   2

    Sample Student Lab Report   10

PART 2   Labs   15

LAB 1   Evaluating Sustainable Practices   16

LAB 2   Analyzing and Interpreting Environmental Data   23

LAB 3   Environmental Science in the Media   32

LAB 4   Sustainability and Business   35

LAB 5   Environmental Site Inspection   38

LAB 6   Urban Ecosystems   43

LAB 7   Experimental Design: Range of Tolerance   50

LAB 8   Experimental Design: Environmental Contamination   56

LAB 9   Landscaping for Energy Conservation   63

LAB 10   Alternative Energy: Wind Power   68

LAB 11 Global Climate Change and Automobiles — 76

LAB 12 Household Contribution to Climate Change — 81

LAB 13 Hydrology and Groundwater Pollution — 86

LAB 14 Stormwater Generation and Management — 93

LAB 15 Applying the Scientific Method: Dowsing for Water — 100

LAB 16 Aquatic Species Diversity and Water Quality — 104

LAB 17 Environmental Forensics — 108

LAB 18 Actual/Virtual Field Trip: Municipal Wastewater Treatment Plant — 114

LAB 19 Actual/Virtual Field Trip: Wetlands and Their Ecosystem Functions — 118

LAB 20 Actual/Virtual Field Trip: Water Treatment Plant — 121

LAB 21 Soil Characterization — 125

LAB 22 Climate Change and Sea Level Rise — 131

LAB 23 Reducing the Generation of Solid Waste — 138

LAB 24 Reducing Campus Food Waste — 145

LAB 25 Compost Facility Planning and Siting — 151

LAB 26 Actual/Virtual Field Trip: Solid Waste Management — 159

LAB 27 Testing the Toxicity of Chemicals — 163

LAB 28 Environmental Risk Perception · 170

LAB 29 Human Survivorship Changes · 174

LAB 30 Indoor Air Quality Inspection · 179

LAB 31 Lung Power · 184

PART 3 Applied Problem Sets · 187

PROBLEM SET 1 The Scientific Method: Observation and Hypotheses · 188

PROBLEM SET 2 The Scientific Method: Results and Discussion · 191

PROBLEM SET 3 Quantification of Environmental Problems · 196

PROBLEM SET 4 Ecosystem Diagram · 201

PROBLEM SET 5 Biogeochemical Concept Map · 202

PROBLEM SET 6 Global Climate Change, CO_2, and You · 204

PROBLEM SET 7 Recognizing Human Impacts · 208

PROBLEM SET 8 Carbon Footprints and Sustainability · 211

PROBLEM SET 9 Oil Consumption and Future Availability · 213

PROBLEM SET 10 Water Quality and Consumer Choice · 215

PROBLEM SET 11 Local Environmental Risk · 218

PROBLEM SET 12 Society and Waste · 220

PROBLEM SET 13   Environmental Modeling   223

PROBLEM SET 14   Environmental Awareness and Ecological Identity   227

PROBLEM SET 15   Trophic Levels and a Tidal Marsh   229

PROBLEM SET 16   Food Efficiency: The Breakfast Assessment   232

PROBLEM SET 17   Life-Cycle Assessment   235

PROBLEM SET 18   Understanding Pesticide Labels   238

PROBLEM SET 19   Review and Reflection   239

PART 4   Appendices   241

    Glossary   242

    The Metric System   248

    Conversion Factors   249

    Numerical Prefixes   250

    About the Authors   251

Introduction

Environmental science is the study of how humans and nonhumans interact with one another and with the nonliving environment. An important element of environmental science is that it views this relationship and interaction as interconnected—a gigantic system. Based on this perspective, environmental science is an interdisciplinary science that integrates concepts from other disciplines, including biology, chemistry, ecology, economics, engineering, ethics, geology, physics, policy science, sociology, and toxicology. Environmental science is an experimental science, meaning that its study is based on the application of the scientific method.

Thus, the labs and problem sets contained in this manual are based on the following:

- Learners need to understand basic concepts associated with the interdisciplinary and interconnected nature of environmental science.

- Learners must understand and experience the scientific method in order to obtain the necessary evidence for environmental problems.

- Learners need to recognize that social science plays a major role in environmental science because environmental problems are socially constructed.

- Learners learn best through doing, through active inquiry.

- Individuals do matter. We all have a personal impact on the environment; we must collectively recognize that personal responsibility is crucial for change.

The labs and problem sets contained in this manual represent the broad spectrum of the interdisciplinary and interconnected nature of environmental science. The subject matter of the labs and problem sets is based on Botkin and Keller's (2014) claim that certain issues are basic to the study of environmental science. These include:

- Rapid human population growth, in conjunction with affluence and technology, is a fundamental cause of environmental problems.

- Human beings affect the environment of the entire planet; therefore, a global perspective on environmental problems must be taken.

- Urban environmental issues and their effects need to be given primary focus.

- Sustaining our environmental resources is crucial for future availability.

- Policy solutions to environmental problems require making value judgments based on knowledge of scientific facts.

This edition has been revised to update the laboratory activities and homework problem sets. We have added several new labs and problem sets but have kept the book at approximately the same size and format in order to keep the cost to the user at a minimum.

These laboratory exercises and problem sets have been designed to use minimum equipment at a minimum cost. This was out of necessity from our own experiences and is in keeping with our views of sustainability and wise resource use. The laboratory and problem sets can be used in a variety of ways; some problem sets may be done as laboratory activities and vice versa. Some problem sets support particular laboratory activities, but there is a great deal of flexibility in how this is done, and we have merely provided general topical subheading in the Table of Contents. Every environmental science instructor has his or her own "tool kit" of favored laboratory activities; some are so essential to understanding environmental science that they have become classics. We have made every attempt to provide original exercises and problem sets while still paying homage to the time-honored classics that appear to belong to the collective whole.

The Internet is used here as a portal to information, but with the appropriate caveats that as a portal it does not discriminate between good or junk evidence. Hence, the learner must become a careful consumer of information and practice *caveat emptor*. And, in fact, that is a key task—we are awash in a sea of environmental information, and we need to sort our way through it. Consequently, we try to use only those Internet websites that appear stable and reliable, and that serve as springboards to further information.

Botkin, D.B., and E.A. Keller. 2014. Environmental Science: Earth as a Living Planet. 9th ed. John Wiley & Sons, New York.

REFERENCE

General Information

PART 1

Laboratory Health and Safety Procedures

The following are general polices and rules for laboratory health and safety in addition to rules specific to your institution:

1. Never smell, taste, or touch an unknown substance.

2. Do not bring food or beverages to the laboratory.

3. Clean up your work area when the lab period is complete.

4. Wash your hands before leaving the laboratory.

5. Keep lab benches and floors free of extraneous books, backpacks, and clothing.

6. Do not use a pipette by mouth at any time.

7. When using microscopes, show the instructor its condition before leaving. Only use lens paper for cleaning lenses.

8. Know the location of the first-aid kit and how to use it.

9. Know the location of the fire extinguisher and/or fire blanket and how to use them.

10. Know the location of eyewash and/or emergency shower stations and how to use them.

11. Know the path to exit the building in an emergency and the designated rally point.

12. Do not apply make-up or other products to your skin in the lab.

13. Wear appropriate protective clothing; do not wear open-toed shoes or sandals.

14. Dispose of materials in proper containers: paper in recycling bins, glass in the special "broken glass" bin, and biohazards in the biohazard bin.

15. Report all chemical spills to the instructor. Spills are handled by the instructor in accordance with university procedures.

16. In accordance with your school's policies, be sure that Institutional Review Board (IRB) or Institutional Animal Care and Use approval is obtained prior to any research or experimentation involving humans or animals.

WRITING LABORATORY REPORTS

A crucial element in science is the ability to effectively communicate your ideas, findings, interpretations, and the experiment's relevance. Writing is not just grammar, but includes content. Writing is part of understanding, and the act of writing is part of the learning process. In this lab manual, we use two primary types of write-ups: formal laboratory reports in the traditional format of the scientist reporting on an experiment and "technical reports;" laboratory and field experience write-ups that reflect particular formats of professional practices. While they may not follow the standard laboratory model of traditional hypothesis testing, technical reports employ a variety of professional experiences and uses from various environmental occupations. Technical reports may include memoranda, inspection forms, and worksheets that are documented with citations and photographs. The formal laboratory reports focus on testing one or more hypotheses. Assume all written lab-related materials must be typed and submitted to the instructor in hard copy unless otherwise instructed. Be sure to number your pages and use appropriate margins. Take the time to do a good job. The time estimates given for completing the labs do not include the time spent outside of class writing up the lab reports. You can expect to spend anywhere from 1 to 3 hours writing up the results of your work.

Before we discuss the style or format of the reports, it is important to consider the background research you will need to do in preparation to perform a study or experiment and to write reports. (Examples and detailed explanations are presented later in this section.) A critical component of environmental science is to identify, locate, and retrieve scientific information—the three-part process of background research. Research is essential to help you find a topic, narrow down the topic, or better understand a chosen topic. A critical step to successful research is to identify and evaluate other studies that have been conducted because their results can assist you in formulating a hypothesis, designing an experiment, or properly framing your approach. For example, if you are going to test a potential toxin on plants, you would want some idea from the literature of what amounts to use so that your experiment does not kill all the plants or, conversely, fail to have any effect at all.

Based on your research, you need to evaluate the validity of the information source you plan to use. Books found in the library traditionally go through an editorial process that involves editors and fact-checkers verifying the information. However, the publishing process for a book might or might not require formal peer review. In contrast, scholarly articles (e.g., academic or journal articles) are generally considered the most reliable because they have undergone peer review. **Peer review** is a process used for checking and verifying the work performed by one's equals—peers—to ensure that it meets specific acceptable criteria, which includes the hypothesis, research design, methods to collect the data, the results, and the interpretation of the results. However, with articles found on the Internet through general searching, peer review is generally not the case. (Exceptions can be found for online versions of peer-reviewed journals and Google Scholar.) Anyone with a computer and access to the Internet can post information or publish a website. Thus, there is no guarantee the information is accurate or true. Therefore, "think like a scientist"—exercise caution and employ skepticism when using information from websites to conduct research. A reference librarian can also help you discern the potential reliability of source material.

You should also be cautious with the age of reference materials. Using a source from 1985 as your primary source for a research paper on the status of global oil production is bound to lead to erroneous conclusions and a poor grade. Similarly, sole reliance on a book published by the American Coalition for Clean Coal Electricity or Greenpeace for your paper on the future of fossil fuels is likely to skew your conclusions due to a bias in perspective. In both of these examples, these sources may serve as appropriate supporting references, but only when combined with more current or neutral-position sources.

As shown in Table 1, it may be helpful to think of sources of information as categories or types, although there may be quite a bit of overlap. For example, popular literature, which is written by journalists, includes a range of reliability. Journalist authors may have scientific backgrounds or they may not. Think of *The New York Times, National Geographic,* and *Science News* as examples of fairly well-regarded popular literature. Professional/Trade publications provide information, case studies, and findings relevant to a specific profession such as water pollution control, waste management, and wetlands mitigation, but generally do not undergo peer review. Gray literature is produced by governments, academics, special commissions, and advocacy groups but is published outside of the traditional publication channels and may not have undergone peer review. These reports may be one-time (e.g., White Papers, commission reports) or periodic such as annual updates or reports. Scholarly literature is written by scientists and other experts in their field. Typically, people who write scholarly literature are primarily researchers and educators, often employed at colleges and universities, who publish as part of their professional obligation to share information and as part of the vetting process of science.

Table 1 Categories of Information

Type	Popular	Professional/trade	Gray	Scholarly
Purpose	Information of a general interest	Information that applies to a specific profession	Specialized or quick distribution and advancing positions or policies	The dissemination of scholarly research
Audience	General public	Practitioners/professionals in a specific field	Decision-makers and researchers	Scholars, researchers, and experts
Examples	• *National Geographic* • *Psychology Today* • *Science News* • *Washington Post* • A Hobbyist's Web page	• Chemical Week • Energy News • Pollution Equipment News • Resource Recycling	• Annual reports to the Legislature • Technical Reports • White Papers • Conference Papers and Presentations	• *Journal of Environmental Management* • *Nature, Climate Change* • *Soil Science Society of America Journal*

Indexing services and packaged library subscriptions (Academic Search Complete, Environment Complete, Web of Science, and others) collect or access journal titles and other reference materials. These collections may be thought of as a filter to ensure information is reliable. As the reasons for why a journal is considered "scholarly" are not always clear or agreed upon, use caution. Indeed, recent years have seen a rise in the variety of "predatory" journals masquerading as scholarly publications, to exploit authors and users (Beall, 2017). Google Scholar is also a type of indexing service but also contains conference papers and gray literature that has not undergone peer review.

FORMAL EXPERIMENT LABORATORY REPORTS	Laboratory reports are your avenue for expressing what you did, why you did it, how you did it, and what you learned in the process. Even if your understanding of the procedures, techniques, and results is perfect and your results are error free, a poorly written report will suggest that you did not thoroughly understand what you have done. Good writing is good writing, be it creative fiction, an editorial, journal article, or scientific communication. Writing reports is not difficult if you remember a few guidelines about writing and the structure of a good lab report.

Your formal experiment laboratory report should have the following components in the following order.

- Title
- Introduction
- Methods
- Results
- Discussion
- References Cited

These components are similar regardless of your field—biology, mathematics, chemistry, physics, engineering, and other science, applied science, and technical fields (STEM)—as they expect a formal structure for experimental laboratory reports. Some stylistic variations may occur among different fields and subfields.

English or metric? Metric measurements (often referred to as SI, or International System of Units) are used in science for presenting facts and figures. Formal experimental lab reports use the metric system. Government, agency, and corporate scientific communications use the metric system, as does all scientific writing for peer-reviewed journals. However, English units are used when presenting information to the general public. Sometimes both are used, as in the case of an environmental inspection report filed with the government—it will have metric units for the scientists and English units, so the public can readily understand it, too. By convention, scientific notation is preferred because it makes the numbers clear. For example, 3.1×10^6 is easier to read than 3,100,000 and similarly, 3.1×10^{-6} is easier to read than 0.0000031. Scientific notation makes it easier to multiply and divide numbers because you add or subtract the exponents. Scientific notation reduces the likelihood of a mistake.

> An excellent resource to help you prepare, organize, interpret, and write your formal laboratory report is LabWrite. LabWrite is an instructional project originating from North Carolina State University; it is sponsored by the National Science Foundation. The LabWrite website is at http://www.ncsu.edu/labwrite.

WHAT DID YOU STUDY?

TITLE

Choose a title that will be helpful to the reader. The report title should be meaningful, concise, and descriptive. Indicate exactly what you studied.

For example, "The Effects of Dissolved Oxygen on Brown Trout (*Salmo trutta*) Mortality" explains the environmental factors manipulated (dissolved oxygen), the parameter measured (survival), and the specific organism used (*S. trutta*).

Always use the scientific (Latin) name for plants, animals, and other biota. Scientific names must be in *italics* or underlined. The genus is capitalized and the species is lowercase. After the first use, scientific names can be shortened by using the first letter of the genus followed by the complete species name. For example, the scientific name for Brown trout is *Salmo trutta* (or Salmo trutta). When referring to the Brown trout again in a lab report, use *S. trutta* (or S. trutta).

To conserve resources, avoid a separate title page and avoid a report cover or binding. If the instructor wants a separate title page, they will likely specify that in a course syllabus.

WHY DID YOU STUDY THIS PHENOMENON?

INTRODUCTION

The introduction should identify the phenomenon you studied/tested and provide relevant background information (e.g., why you studied it, its environmental relevance). Generally, you should include information from other studies or documents, which must be properly cited. The background component of your introduction should reference pertinent literature to place your experiment in the context of past and present research efforts. The background research in your introduction will increase the likelihood that your experiment will cover a range that produces meaningful results. For example, if you are testing plant toxicity and salt, you want to select salt concentrations that will affect your plants, so you need some background on the plants and on salt to help you select an appropriate concentration. Your introduction might have a statement like this:

According to McKenzie (2000), white pine (*Pinus strobus*) is highly vulnerable to salt.

(And be sure you have McKenzie (2000) properly listed in your References Cited section.)

The introduction builds toward your hypothesis. The hypothesis is your educated prediction about the topic, which can be tested. Experiments generally look for cause-and-effect relationships, which allow you to predict what will happen if something is changed or something occurs. The factors changed in an experiment are called variables. Thus, your hypothesis must contain the variables to be tested. The independent variable is the variable you change ("the cause"), the dependent variable is the variable you monitor to see if the changed independent variable made a difference ("the effect"), and the control variable is the variable you want to be constant.

In the example below, we use **bold** and <u>underline</u> to demarcate the variables. The experimental manipulation is the <u>independent variable</u> and the resultant effect is the **dependent** variable. Think of it as testing your proposal that a dependent variable such as **seed growth** (dependent variable) depends on <u>pH of precipitation</u> (independent variable).

Our example:

Hypothesis: <u>Precipitation with a pH of 3.6</u> significantly reduces the success of red maple (*Acer rubrum*) **seed germination**.

An alternative hypothesis helps serve as a check on the experiment. Thus, a null hypothesis might be included; this is when you postulate that the experiment (<u>independent</u> variable) will have no effect on your test subjects (application of a **dependent** variable). Thus, "red maple (*Acer rubrum*) **seed germination** is not affected by <u>precipitation with a pH of 3.6</u>."

METHODS

WHAT DID YOU DO? HOW DID YOU DO IT?

In the Methods section of a formal lab report, you will describe how and when you did your work, including experimental design, experimental apparatus/equipment, methods of collecting and analyzing data, sample size, and types of experimental control. This section should be written in the past tense because you have already done the experiment. This section must include complete details and be written clearly enough to allow readers to duplicate the experiment; thus, it must be written in chronological order using complete sentences and not bulleted or numbered lists. It should not be written in the form of instructions or as a list of materials as in a laboratory manual. Instead, use a narrative, which describes in the active voice what you did. For example, *We filled six petri dishes with 20 mL of tap water in each*. Be sure to use metric measurements, scientific notation, and label all units. Would an illustration (e.g., figure or photograph) help the reader to understand how you did the experiment? Will your methods allow you to obtain enough data to support a statistical analysis? If you are testing something such as toxicity, have you selected levels of treatment that will provide a range of results?

RESULTS

WHAT DID YOU FIND?

In the Results section, you present your observations and data with *no* interpretations or conclusions about what they mean. Tables and graphs should be used to supplement the text and to present the data in a synthesized, understandable form. Use the past tense to describe your results. Provide statistics to summarize and describe the data as appropriate.

When using tables or figures, each table or figure must be introduced within the text, and not placed in an appendix. Although some formats call for placing the tables and figures after the references and before the appendix, unless your instructor says otherwise, the best place for the figures and tables is on or near the page in the text where they are referenced. All tables and figures must be numbered and have self-explanatory titles so that the reader can understand their content without the text (e.g., Table 1—Percentage of Red Maple (*Acer rubrum*) Seedlings Exhibiting Visible Injury After Exposure to Water with a pH of 3.6). Assign numbers to tables and figures in the order that they are mentioned in the text. Tables and figures are numbered independently of each other (i.e., Table 1 and Table 2, and then Figure 1 and Figure 2). Tables are labeled at the top and figures at the bottom, and both have captions at the bottom. Tables display data (including words) in row or column format. All other items (graphs, photographs, drawings, diagrams, maps, etc.) are referred to as figures. Axes on figures should be labeled. If you have conducted statistical tests and placed the results in figures, provide means and p-values. The text should support the tables and figures by pointing out things such as trends, important differences, and unexpected results. This helps orient the reader to the points that you will be interpreting in the Discussion section. Many writing manuals and other references provide detailed information on the use of tables and figures (e.g., Hacker and Sommers, 2015; McMillan, 2017). Many colleges and universities have posted their own recommendations and guidelines online.

WHAT DOES IT MEAN?

The previous section reported your data; now you explain what you think your data mean. Describe patterns and relationships that emerged. Focus on the most important or probable explanations for the results. Explain how any changes to, or problems with, the experimental design/procedure may have affected the results. Without making excuses, what are the limitations of the study? Tie your interpretations to the original hypothesis. Rejecting a hypothesis is just as valid as accepting it. Explain what you learned despite the limitations of the data set. What would you recommend for future studies?

Use a standard format to cite any literature used in your report (e.g., your textbook, journal articles, Web pages, books). Sources must be credited if you obtain ideas or thoughts from them, even if you are not giving a direct quotation (e.g., Botkin and Keller, 2014). In the text of your report, if you cite specific information, or quote data or persons, cite references using the author's surname, the year of publication, and the page number(s) for the source quoted or paraphrased. Failure to credit sources leads to the academic and ethical sin of plagiarism. In addition to citing a source, you need to be sure that it is the proper source. The Internet is a portal to information, but it tells you nothing about the credibility of the information. Reference librarians are an underutilized source for credible sources. Some Internet sites, such as Wikipedia, are open-edited and therefore not generally appropriate as references for scientific writing.

No single, correct citation format exists; each discipline (e.g., biology, economics) and each journal has a preferred citation format. Thus, what is presented below is one particular style, but you will see other styles in your research. It can be frustrating do I capitalize the first letter of each word in a journal article title? Do I put the title of the journal or website or book in italics? Better to regard these questions as a matter of style rather than to omit valuable information. You can use our style guide and keep in mind that the most important aspect is to be consistent and to ensure that all necessary information is contained in the citation.

Some students who take an introductory environmental course come from the humanities and liberal arts fields, and therefore may be accustomed to the MLA (Modern Language Association) citation format. It is appropriate for the liberal arts and humanities but is not used in the STEM fields. One of the reasons is that MLA just uses the author name for in-text citations and science writing should have the author and *year* of publication. The MLA and typical STEM formats like APA both require page numbers for quotations and paraphrases. In science, more information is required; we need the year and the page number. The year is important due to the changes in science and the accountability of sources.

For the purposes of this lab book, and if not otherwise specified by your instructor, you might choose to write your assignments and reports with the following citation format from the Soil Science Society of America Journal (a typical science format).[1] If you are using EndNote® (http://www.endnote.com) to create your reference list or bibliography, use "Ecology" as your output style. Your instructor may recommend a different citation format, but whatever you use, the idea is to be consistently and systematically accountable for where your information comes from. You may find that the reference does not always nicely fall into sortable categories. For example, a periodical publication might not give the day and month, just the year. But you can generally get it sorted if you remember the idea of accountability and give the reader as much information as needed to track down the reference.

---

[1] Source: American Society of Agronomy, Crop Science Society of America, Soil Science Society of America. 1998. Publications Handbook & Style Manual. ASA, CSSA, SSSA, Madison, WI.

EXAMPLES OF PROPER REFERENCE CITATIONS

JOURNAL ARTICLES OR PERIODICAL REFERENCE

Single author

Beall, J. 2017. What I Learned from Predatory Publishers. *Biochemia Medica* 27:2:273-9. Available online at http://www.biochemia-medica.com/2017/2/273. (Verified 26 June 2017).

Connell, J.L. 1974. Species Diversity in Tropical Coral Reefs. *Science* 234:23–26.

McKenzie, R. 2000. Right Tree – Right Place: White Pine and Salt Tolerance. Purdue University Extension Service. FNR-FAQ-10-W. [Online] 2000. Available at https://www.extension.purdue.edu/extmedia/fnr/fnr-faq-10-w.pdf (verified 19 June 2017).

Multiple authors

O'Rourke, D., L. Connelly, and C.P. Koshland. 1996. Industrial Ecology: A Critical Review. *International Journal of Environment and Pollution* 6:89–112.

BOOKS

Single author

McHarg, I.L. 1971. *Design with Nature*. Doubleday, Garden City, NY.

McMillan, V.E. 2017. *Writing Papers in the Biological Sciences*. 6th Ed. Bedford/St. Martin's, Boston, MA.

Multiple authors

American Society of Agronomy, Crop Science Society of America, Soil Science Society of America. 1998. Publications Handbook & Style Manual. ASA, CSSA, SSSA, Madison, WI.

Botkin, D.B., and E.A. Keller. 2014. *Environmental Science: Earth as a Living Planet*. 9th ed. John Wiley & Sons, New York.

Hacker, D. and N. Sommers. 2015. *A Writer's Reference*. 8th ed. Bedford/St. Martin's, Boston, MA

Chapter in a book

Rabe, G.B. 1999. Sustainability in a Regional Context: The Case of the Great Lakes Basin. pp. 248–281. *In* D.A. Mazmanian and M.E. Kraft (eds.) *Towards Sustainable Communities: Transition and Transformations in Environmental Policy*. MIT Press, Cambridge, MA.

Internet

Citations for Internet (Web) sites should be similar to print media citations, including author, publication date, article title, site title, URL, and date that the information was posted (or when the address was accessed). Many references can be in both print and online formats.

Internet publication and source

Krupnick, A. J., I Echarte, L. Zachary, and D. Raimi. 2017. WHIMBY (What's Happening in My Backyard?): A Community Risk-Benefit Matric of Unconventional Gas and Oil Development. Resources for the Future. Final Report [Online]. June 2017. Available at http://www.rff.org/research/publications/whimby-what-s-happening-my-backyard-community-risk-benefit-matrix (verified 22 June 2017).

Tucson Audubon Society. 2017. Tucson Audubon Society Southeast Arizona Rare Bird Alert. Available at http://tucsonaudubon.org/go-birding/southeast-arizona-rare-bird-alert (verified 19 June 2017).

Government website

U.S. Environmental Protection Agency (US EPA). 2017. Home Page [Online]. Available at http://www.epa.gov (verified 19 June 2017).

Some of the laboratory exercises are not formal experiments and thus are not written up as an experimental laboratory report. These lab exercises contain a series of questions designed to promote a particular experience or a range of exploration for an environmental issue, or they represent an approximation of the type of professional reports done in various environmentally related occupations. Some labs require you to respond merely to the questions, whereas other labs require you to use the questions to guide the development of your response in narrative form or in a memorandum. As a guide, all the labs provide suggested headings for your technical report. Be sure that your answers are complete, and when appropriate, show your work (how you did your calculations). When answering questions, be sure to write out full answers. Never simply state *No* or *Yes*, but define yourself or explain why a No or Yes response is appropriate. Provide complete references for any information that comes from other sources, and include page numbers if quoted or paraphrased. (In essence, your instructor should be able to quickly and easily locate the exact source with the information you provide.) These citations should be provided in the same format and manner as for formal experimental laboratory reports.

> **TECHNICAL REPORTS**

Number the pages (begin with the second page only if you have a separate title page; a separate title page itself does not get a page number put on it). Numbering helps readers find their way through your product. Use a writer's guide (e.g., Hacker, D. and N. Sommers. 2015. *A Writer's Reference*. 8th ed. Bedford/St. Martin's, Boston, MA. or McMillan, V.E. 2017. *Writing Papers in the Biological Sciences*. 6th ed. Bedford/St. Martin's, Boston, MA) as a reference for proper writing. These books contain much good advice.

> **FOR ALL WORK**

Do not use contractions. This has the added benefit of avoiding confusion between homophones like *it*'s and *its, your* and *you're*. Avoid anthropomorphisms (e.g., *cicadas feel scared in the presence of hungry birds*), as plants and animals are not people. Use topic sentences to begin paragraphs. Be sure that you use complete sentences. Use active voice. Although we are using the second person (you) point of view in writing this textbook, you should not do that when you write your formal or informal lab reports (it is okay for a lab book, but not a lab report!). Otherwise your report could sound "pedantic"; not the way to impress an instructor or supervisor.

Use first person (*I, we*) or third person (*he, she, it, one,* and *they*). If you are not sure about a word and it is not in our glossary, just look it up in the dictionary. Watch for common mistakes such as confusing *cite* (to quote) with *site* (a place) and effect (usually a noun) and affect (a verb). Use scientific notation and write the units for numbers. All formal lab reports should use the metric system. Assume technical reports use the metric system too, although sometimes the English system is given as well. An exception to the metric in the United States: it is generally acceptable for environmental inspection forms and other documents intended for public use (e.g., something that might be given to a homeowner) to only use English units. If in doubt, use both units, or check with your instructor.

> **AFFECT VERSUS EFFECT**
>
> In general, affect is usually a verb and effect is usually a noun.
>
> **Affect** means "to influence" something (e.g., The salt affected the germination of the plant.)
>
> **Effect** generally means "a result" of something (e.g., The salt had an effect on the germination of the plant.)

Your instructor may have examples of the required format for laboratory and technical reports in your course, and of the type of research and critical thinking to be displayed in them. However, to give you some idea of what might be done, an example, with comments, is provided in the next section.

Sample Student Lab Report

This section contains an example of a formal experimental laboratory report submitted as an assignment using the previous edition of this laboratory manual. Below we have provided comments on the lab report. Occasionally we use strikeouts to indicate some words we recommend deleting as part of editing. We point out some positive features of the report in addition to some things we recommend fixing. You can expect some instructors to mark-up assignments far more than others, but you should not expect them to edit your report for you. Most instructors will have their own areas of emphasis, style, and expectations for reports and assignments; be sure to find out what they are.

Name
Course
Professor
Lab #
Lab partners
Date

Environmental Contamination:
Toxicity of Purple Flame Ice Melt (sodium chloride) on Growth of *Triticum aestivum*

Capitalize this since it is in a title

I. INTRODUCTION

The primary component of most road salts is sodium chloride generally consisting of 40% sodium ions and 60% chloride ions (NH Department of Environmental Services, 2016). Chloride ions easily dissolve in snow and ice melt, thus impacting roadside vegetation and nearby bodies of water (NH Department of Environmental Services, 2016). The New Hampshire Department of Environmental Services (2016) noted that from 2008 to 2010, the number of bodies of water damaged by chloride ions increased from 19 to 40 as road salt use continued to increase. As a de-hydrant, salt causes an osmotic reaction when it comes into contact with plants (and other biotic beings). Dihydrogen oxide (water) is forced out of the cells in order to mitigate and dilute the external hypertonic concentration of salt. By removing the concentration gradient, the plants ultimately reach equilibrium. Salt therefore draws water out of the cells and dehydrates the plant. Plants can thus suffer from nutrient deficiency as salt causes damage to the leaves and roots.

Good opening sentence

This explains the relevance of the problem

From 1950 to 1960, the annual use and application of road salt in the United States increased from one to ten million tons. On highways, salt application varies from 200 to 400 lb/lane-mi (Road Salt Use in the United States – TRB, 1990). Interstate 95 in Maine, which is approximately 303 miles long and is primarily two lanes, is covered by a minimum of 133,200 pounds of salt annually, using the lowest average salt application of 200 lb/lane-mi.

Reference citation is (author, year), and add page number if quote or paraphrase (Author, year: page).

Four lanes

Salt is also used by homeowners for driveways and sidewalks. One popular commercial brand is Purple Flame™ Commercial Ice Melt. According to the Salt Depot Material Data Safety Sheet, Purple Flame™ Commercial Ice Melt consists of Sodium Chloride (92%), magnesium chloride (6%), calcium chloride (1%) and potassium (1%). Additionally, Purple Flame gets its name from the acid purple liquid colorant blend of xanthene and triphenylmethane dyes. Overall, the ice melt is relatively nontoxic to humans, causing mainly subtle irritation when coming into contact with skin or eyes, and possible nausea and vomiting if ingested. The LD50, as stated

Because a source is being referenced, you need to put the year in parentheses, in this case (2014).

Chemical names are not proper nouns thus they should not be capitalized.

10

by the Material Safety Data Sheet, for oral ingestion by rats is 3000 mg/kg (Salt Depot, 2014). While no information could be found, the toxicity to plants is an important question given the predominate flow of salt-contaminated ice melt onto vegetated roadsides.

In this experiment, *T. aestivum* will be studied under varying road salt (Purple Flame) amounts: 0.1, 0.2, 0.4, 0.8, 1.6 grams. It is hypothesized that *T. aestivum* seeds exposed to a concentration of 0.4 grams or greater of Purple Flame Ice Melt will be a phytotoxic level that will prevent plant growth and maturation.

H_0: 0.4 grams of Purple Flame Ice Melt will not be phytotoxic to *Triticum aestivum*.

When using the formal genus and species name for an organism, the name must be italicized or underlined. Within a paragraph, the genus only needs to be spelled out entirely the first time it is used. After that, you can abbreviate it with its first letter. For example, if you have already used the term Triticum aestivum in a paragraph, the next time you use it, you can write T. aestivum.

II. METHODS

The largest concern for this experiment is the proper selection of Ice Melt used on the seeds. Sodium chloride is toxic to plant growth; however, the concentration at which the lethal dose for plants occurs was difficult to pinpoint. Ingestion of three grams of sodium chloride by rats is enough to cause acute oral toxicity. The average biomass of rats is approximately 240 grams, whereas the average biomass of ten *T. aestivum* is only 0.23 grams. The seeds have roughly one thousandth the biomass of a rat, therefore, lethal dose of sodium chloride by plants could be reached at perhaps 0.69 grams of NaCl (not in solution). Due to restrictions from the scale (only to a tenth of a gram), the experimental trials began at 0.1 grams and doubled with succeeding each plant. The amount of salt in each part of the trials, if completely dissolved could possibly prevent all plant growth. The idea, however, is that by adding crystallized salt, and not adding a salt solution to the soil, only a small portion of the Ice Melt will dissolve and penetrate the seeds. Additionally, cold water (~10°C – 15°C) was added to the soil, thus limiting the amount of possible deliquesced sodium chloride.

This is an example of a good transition sentence that connects the background information on the study problem at large and what the student's study is about.

switch order

Eighteen plant containers were used for a total of three trials with five different concentrations of ice melt and three controls. Twenty-four grams of organic soil were weighed and placed in each of the eighteen 5-ounce containers to maintain consistent environmental conditions for each container. The tray was divided with masking tape into three rows to separate each trial, and an offset column to differentiate the control group from the treatment groups. Each plant container was placed in the tray. Ten seeds were counted and then placed into each container at about the depth of one knuckle. The seeds were massed for later comparisons between the LD50 of the oral consumption of salt by rats. Moving from left to right (and skipping the control group), 0.1g, 0.2g, 0.4g, 0.8g, 1.6g of Purple Flame ice melt was placed in each trial container (1-5) respectively for three trials. Twenty milliliters of tap water were poured into each plant container, and the tray was thereafter loosely covered with plastic wrap to maintain a constant humidity for the plants. Ambient conditions such as light, temperature, humidity, soil nutrients, seeds, and water could all change the outcome of the experiment if not controlled. To limit these variables, all plants were giving the same amount of organic soil, watered the same amount and placed under UV bulbs to offer a consistent and unwavering light source. Humidity was controlled using plastic wrap loosely placed over all the plants in order to maintain the interior conditions of the environment and act as a green house. Slight changes in environment could enhance or hinder the growth of plants, therefore all plants were monitored under identical conditions over the same time period.

Good, you are using past tense, which is appropriate since the experiment was completed

How did you decide this amount?

given

LED lights

greenhouse is one word

As shown in Figure 1, the tray was then placed under LED lights for uniform and continuous light. The seeds were watered with 10mL of tap water every two to three days until the end of the study's growth period. The plants were harvested and and then the plant heights in each cup were recorded and measured in centimeters with a ruler to determine plant growth on a numerical scale. For each trial, the mean plant growth was compared to the treatment on each respective plant (i.e., 1Ct to 1a, 1Ct to 1b and so on).

Delete 2nd "and"

How tall the plants were?

FIGURE 1 Study apparatus showing the three trials

III. RESULTS

Figure 2 depicts the growth of the *T. aestivum plants* over a one-week period. Each container, moving from left to right, is contaminated with a specified amount of salt, doubling from 0.1 grams to 0.2 grams to 0.4 grams and so on. The first container is the control, which was no exposed to any contamination. The center container, containing 0.4 grams of Ice Melt, experienced severe stunted growth.

No should be not

By doubling the salt contamination values with each proceeding container of seeds (five different salt values and a control per trial), the plants showed a consistent decline in growth. With only 0.1 gram of salt, the *T. aestivum* showed a slight decline of growth, which was then followed by a large decline with 0.2 grams and 0.4 grams of salt added. Little to no growth was observed passed 0.4 grams. As shown in Table 1 and Figure 3, the control average of 12.7325 centimeters was reduced to 6.43 centimeters with a salt contamination of 0.4 grams. This is a reduction by almost half.

Limit this to 2 decimal places = 12.73

FIGURE 2 Tray of *T. aestivum* after one week of growth

Table 1   Growth of *T. Aestivum* with Varying Concentrations of Salt.

Always provide the units. Otherwise the data are ambiguous.

	Control	0.1 g	0.2 g	0.4 g	0.8 g	1.6 g
Average Growth	12.7	10.51	6.94	6.43	0.177	0.00

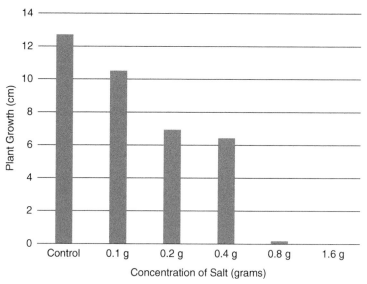

FIGURE 3 Mean growth rates of *T. aestivum* by salt concentration.

Table 2 Statistical Results of the Three Trials.

Amount of Salt	Mean Growth	*p*-Value
Control	12.7 cm	-
0.1	10.51 cm	0.396
0.2	6.94 cm	0.05
0.3	6.43 cm	<0.01
0.4	0.17 cm	<0.01
0.5	0.01 cm	<0.01

Be sure to specify the units.

A series of *t*-tests were conducted to statistical analyze the results to determine if the hypothesis could be accepted. The results of the *t*-tests are presented in Table 2.

With an average *p*-value of 0.396, the observed growth of red winter wheat plants was not significantly statistically affected by road salt contamination of 0.1 grams. The *T. aestivum* contaminated exposed to 0.2 grams of Purple Flame Ice Melt had a p-value of 0.05 or less (0.49 to 1.05E-19 for NaCl values greater than 0.1).

Consider rewording this sentence to make it even more clear (say it out loud to yourself to see how it flows).

IV. DISCUSSION

Concentrations of 0.2 grams and greater of ice melt resulted in *p*-values of 0.05 or less. This value indicates that there is at most a 1 in 20 chance that the results regarding the differential growth of the plants occurred by chance. Looking at the average growth of plants in each trial, ~~it is shown that~~ at 0.4 grams of Ice Melt contamination, the plants experienced stunted growth by almost half. With an average *p*-value of 0.00045, ~~it is evidenced that~~ the stunned growth of the red winter wheat plants had a 0.045% possibility of occurring by chance. This indicates that the reduced plant growth likely occurred from salt contamination. Conclusively, 0.4 gram contamination of crystallized Purple Flame Ice Melt was phytotoxic to *T.* aestivum.

Replace with "indicates"

change stunned to stunted

Considering that the plants experienced stunted growth at only 0.4 grams of salt exposure, the discussion of the effects of road salt become broader. With a minimum approximation of 133,200 pounds of salt and a maximum estimation of 266,400 pounds of salt used on Interstate

95 in Maine, it is evident that most of this product ends up on the side of the road and as run off into the road-side vegetation and water supply. The over application of salt, meant to keep busy roads usable during winter months, causes a detrimental effect to the nearby vegetation. As a dehydrate, only 0.4 grams of salt is toxic to plant growth, stripping vital nutrients and water from *T. aestivum* through the biological process of osmosis.

Nice point—provides context.

Vegetation additionally protects soil erosion. Grass (although not as beneficial as shrubland or forest) keeps soil in place with a complex root system. Without vital topsoil, plants cannot grow due to lack of nutrients. In the US, topsoil is being lost at a rate that is ten times faster than it is being replenished (Powers, 2016). Soil loss is detrimental to agriculture and natural plant growth as it removes the vital nutrients required to provide sustainable and healthy vegetation.

these need to be referenced

Conclude with a discussion of how you might refine or otherwise improve this experiment and of further experiments that could be done.

Excess salt creates a toxic soil environment known as soil salinity, which can prevent plant growth. Ions in solution from compounds in various industrial ice melts (NaCl, $MgCl_2$, etc) can cause ion toxicity in plants. As evidenced by the stunted and completely halted red winter wheat growth with salt concentrations of 0.4 grams and greater, excess ions (leached into the soil from NaCl) cause plants to become toxic. Growth is therefore limited if not completely prevented. Vast concentrations of salt and ice melt are introduced to plants on a yearly basis. Plant death, soil salinity and excess soil erosion are at a greater risk of occurring due to our uncontrolled application of road salt and ice melt.

V. REFERENCES

References Cited should be in alphabetical order by author.

Home depot ice melt pellets. (n.d.). Retrieved November 03, 2016, from http://dompetlufas.com/home_depot_ice_melt_pellets

Author first. Capitalize Depot.

Material Safety Data Sheet - Sodium Chloride. (2013, May 21). Retrieved November 3, 2016, from https://www.sciencelab.com/msds.php?msdsId=9927593

NH Department of Environmental Services, Water Quality Impacts - Environmental, Health and Economic Impacts of Road Salt – Salt Reduction - Watershed Assistance Section – NH

Department of Environmental Services. (2016). Retrieved November 03, 2016, from http://des.nh.gov/organization/divisions/water/wmb/was/salt-reductioninitiative/impacts.htm

Be sure to provide the URL, which is an essential component of a correct citation.

Powers, M. (2016, October 3). Soil, Science, and Society: We're Running out of Dirt. Retrieved November 10, 2016, from http://www.fewresources.org/soil-science-and-society-were running-out-of-dirt.html

Road Salt Use in the United States - TRB. *Online Publications*. Online Publication, 1990. Retrieved November 10, 2016.

Avoid all –capitals; just first letter of each word.

SAFETY DATA SHEET (SDS) - sfm.state.or.us. (2014, July 15). Retrieved November 3, 2016, From http://www.sfm.state.or.us/CR2K_SubDB/MSDS/PURPLE_HEAT_ICE_MELTER.PDF

Labs

PART TWO

Evaluating Sustainable Practices

OBJECTIVES

- Identify and evaluate specific examples of sustainable practices of a site (e.g., campus, community, institution, shopping mall).

- Prepare a technical report.

KEY CONCEPTS AND TERMS

- ✓ Alternative energy
- ✓ Anthropogenic
- ✓ Energy conservation
- ✓ Energy efficiency
- ✓ LEED, Leadership in Energy and Environmental Design

- ✓ Pollution prevention
- ✓ Potable water
- ✓ REC, Renewable Energy Credits
- ✓ Source reduction
- ✓ Sustainability

TIME REQUIRED:[1] 1.5 hours in the field and 0.5 hour in the library or site's sustainability office.

INTRODUCTION

Sustainability, or sustainable development, is generally referred to as meeting the needs of the present without compromising the ability of future generations to meet their own needs (WCED, 1987). The concept of sustainability is built on three pillars: Society, Economy, and Environment. A sustainable society is based on a green economy with a fair distribution of resources, improved human well-being and social equity, and reduced environmental impacts.

How does this apply specifically to a site such as a campus, your community, a building, or a shopping mall? A "site" can adopt and implement sustainable actions that serve not only the planet locally and globally but also the internal community including students, staff, faculty, visitors, workers, residents, customers, and the external community such as neighbors. Sustainable actions can include innovative solid waste management practices, implementing energy efficiency and conservation, conserving water, reducing food waste, and using sustainable materials in buildings. A guiding framework for implementing sustainability focusing on reducing environmental impacts at a site is pollution prevention. **Pollution prevention** is any practice that reduces, eliminates, or prevents *pollution* or waste at its source, also known as "**source reduction**." Source reduction is more desirable than recycling, treatment, and disposal because the last three focus on managing waste and pollution *after it is created*. Successful pollution prevention reduces environmental impacts and enhances the protection of workers' health. It also saves money as fewer materials or products are purchased, and when less pollution or waste is created, there is a corresponding decrease in management or treatment costs.

SOURCE REDUCTION

Source reduction seeks to reduce the quantity and/or toxicity of pollutants and waste (before they are created) through a variety of approaches:

- Using less harmful materials
- Improving production processes to reduce creation of waste
- Reusing materials
- Preventing spills
- Implementing water and energy conservation and efficiency

---

[1] Time estimates for field and in class/library research time, not the time spent writing up the report.

- Camera
- Clipboard
- Internet access
- Thermometer, to measure ambient conditions

In this lab, you will be identifying and evaluating sustainable and nonsustainable practices and features at a site. The final outcome of this lab is to produce a technical report, which you will title "Sustainability Evaluation Report of (name of site)." Because technical reports are a professional-level type report often prepared by environmental and sustainability professionals for clients, it is critical to properly cite your sources. Also, technical reports almost always contain photographs to document positive and negative examples. When you include photographs, be sure to label them with date, time, and specific location.

Below are the various components of sustainability you will be trying to locate and assess. Note that not all of these will be relevant to your site and your site may have components not listed below, but should be considered. In the appendix, there is a Sustainability Site Evaluation Checklist that you should use to guide your site inspection and evaluation. Be sure to attach a copy of your completed checklist as an appendix to your technical report.

1. SUSTAINABILITY POLICY

Organizational policies or mission statements are important to provide guidance on operations and decision-making. Locate the sustainability policy and/or mission statement for your site and assess the following (be sure to properly cite everything—see the checklist in the appendix):

A. What is the policy or mission statement related to sustainability (if there is no such policy or statement, note this)?

B. How easily accessible is the policy or mission statement (assess this as an external community member)?

C. How clear, easy to understand, and measureable is the policy or mission statement?

D. Is there an Office of Sustainability or a Sustainability Coordinator (or something similar)? What is its name or title?

E. Is there a sustainability website to assist internal and external community members? If so, what tools are available on the website to assist users in understanding, assessing, and adopting sustainable practices?

2. BUILDINGS

The typical person spends some 84% of their time inside buildings. Thus, the environmental impact of buildings includes impact on our health. A relatively new field "green building" is increasing in importance. Green building (sometimes called sustainable building) is the practice of creating and incorporating healthier and more resource-efficient and less ecological impactful models of construction, renovation, operation, maintenance, and demolition of all types of buildings.

The U.S. Green Building Council created and maintains **LEED (Leadership in Energy and Environmental Design)**, which has a series of four ratings (certified, silver, gold, and platinum) for the design, construction, operation, and maintenance of green buildings.

A. Which, if any, buildings on your site are LEED certified? Where is it located? What is the primary function of the building?

B. What is the rating status level of the building?

C. Does the site have a policy regarding future buildings and LEED certification?

D. If there are no LEED buildings, what recommendations can you offer for the most viable building as a potential candidate for certification?

3. ENERGY EFFICIENCY AND CONSERVATION

The generation and consumption of energy are the primary activities responsible for anthropogenic (**anthropogenic** means human-caused or created) carbon emissions. There are two basic strategies to improve the sustainability of our energy consumption, conservation, and efficiency, which apply to every activity that consumes energy. **Energy conservation** is any behavior or technology that results in the use of less energy (e.g., turning lights off when leaving the room, installing motion-detector-based lighting, or using a sleep function on electronics). **Energy efficiency** is use of technology that requires less energy to perform the same function (e.g., using LED lights that require less energy instead of fluorescent or incandescent lights, driving hybrid cars, using a clothes line to dry clothes). With electrical appliances, an easy approach to assess if they are energy efficient is to determine if they are Energy Star™ compliant, a voluntary labeling system for energy-efficient appliances.

Examine a building selected by your instructor:

A. Examining the outside of the building, how many individual *window-based air conditioners* are installed?

B. Locate a few unoccupied rooms. Using your thermometer, record the temperature (you are assessing whether unused rooms are cooled or heated) and note the size (square footage) of the rooms.

C. Locate some vending machines. How many of them are lighted? How many of them are refrigerated? (Do any of these machines use Vending Misers™? You may have to search for an explanation of this device.)

D. Identify and describe (room number if available) two specific examples of excessive/extraneous indoor lighting. This can be lights left on, too many lights, plentiful natural light but with electric lights on, etc.

E. If it is still daylight, find and describe three examples of unnecessary outdoor lighting (provide specific locations).

F. If there are laundry facilities, are washers and dryers labeled as Energy Star™?

4. ALTERNATIVE ENERGY

While the above sections discussed how to consume energy more sustainably, another sustainable action is to produce energy that emits no or less carbon, which is generally referred to as **alternative energy** because it is an alternative to traditional fossil-fuel-generated energy.

A. Does the site have solar power installed? If yes, be sure to identify the type of solar power installed. (There are two major types of solar energy: Photovoltaic technology directly converts sunlight into electricity, whereas thermal technology harnesses solar heat generally to heat water.)

B. Are there other types of alternative energy installed (e.g., wind, biomass, geothermal)?

C. Does the site purchase **RECs (Renewable Energy Credits)** to offset carbon emissions? (RECs are purchased certificates that represent one megawatt-hour of electricity that is generated and delivered to the electricity grid from a renewable energy resource.)

5. WATER CONSERVATION

Conserving water is not only important because water is a scarce resource, especially in areas of low precipitation, but the delivery of agricultural, industrial, and potable water (**potable** means water suitable for drinking) consumes energy and requires significant infrastructure. Thus, the less water consumed the lower the energy and infrastructure costs.

Examine a building selected by your instructor:

A. Are there water conservation actions used for the landscaping (e.g., native plants, minimal turf grass, drip irrigation)? If so, which ones have been adopted?

B. How many low-flow sink faucets are installed in the building? What percent of the total number in the building are they?

C. How many waterless or low-flow toilets are installed? What percent of the total are they?

6. SOLID WASTE MANAGEMENT

Solid waste refers to any waste, regardless of whether it is a solid or liquid, generated by residential and institutional facilities and operations such as homes, dorms, campuses, and buildings. Solid waste includes paper, plastics, food, textiles, metal, and glass. As discussed earlier, implementing pollution prevention through source reduction reduces the creation of waste. However, when created, managing solid waste by reusing and recycling recovers materials and reduces energy consumption as opposed to disposal through landfilling or incineration.

Examine a building selected by your instructor:

A. How many water-bottle-filling stations are there? If the stations record bottles saved, how many plastic bottles have been saved with each water-bottle-filling station?

B. Do any food kiosks or stores on site provide single-use plastic bags? Plastic straws? Expanded polystyrene containers for takeout food?

C. How many recycling stations are there? To what degree are the recycling stations conveniently close to a major solid waste generation point (e.g., food kiosk)?

D. Near the stations, how many specific examples of visual or written prompts (signs, labels, etc.) are there that promote sustainable behavior for prevention of waste generation? Reuse? Recycling?

These relate to the site and not a specific building:

E. Is there a "move-out" collection and reuse program for the dorms?

F. Is there a surplus or reuse center (e.g., for furniture, electronics, binders) on the site?

7. FOOD

A. At a food kiosk or cafeteria, how many examples of local food can you find (list them and make sure they are local)?

B. How many written prompts (signs, labels, etc.) that advertise local food options are there?

C. Is the cafeteria trayless?

D. How many written prompts (signs, labels, etc.) are there that encourage customers to not waste food?

E. Is food waste generated on the site (e.g., cafeteria, food kiosks, snack bars) also composted on-site? Is it composted off-site?

F. Are there any community gardens on the site?

8. ALTERNATIVE TRANSPORTATION

A. How many bike racks are installed? Are there enough installed? Are they convenient for users?

B. Is there a bike share program?

C. How many carpool parking lots or spaces are there?

D. Is there a car share or ride share program on the site?

E. Is the site served by public transportation? How convenient is it to use?

F. How many charging stations are there for electric vehicles? Where are they located?

9. OTHERS

Identify any other actions or practices you observed that promotes or reduces sustainability.

10. OVERALL ASSESSMENT

Provide a summary opinion of whether or not you would define your site as sustainable by relating to specific instances of sustainable and nonsustainable actions you found during your inspection. Be sure to include specific recommendations on how the site can become more sustainable.

| **WRITE-UP** | It is now time to prepare your technical report, use the following headings for your report: Sustainability Evaluation Report of (name of site). |

    I. Sustainability Policy

    II. Buildings

    III. Energy Efficiency and Conservation

    IV. Alternative Energy

    V. Water Conservation

    VI. Solid Waste Management

    VII. Food

    VIII. Alternative Transportation

    IX. Other

    X. Overall Assessment

| **REFERENCES** | WCED. 1987. Our Common Future, World Commission on Environment and Development (WCED). Oxford University Press, Oxford, UK. |

APPENDIX

Sustainability Site Evaluation Checklist

The purpose of this checklist is to help guide on your site evaluation, but be sure to answer all the questions, showing documentation and support as appropriate.

Item	Reply	Documentation, Notes, Photographs Taken
Sustainability Policy		
Is there a sustainable policy or mission statement?		
Is the policy/statement accessible?		
Is it understandable, measureable?		
Is there an Office of Sustainability or a Sustainability Coordinator?		
Is there a sustainability website? Does it contain sustainability tools?		
Buildings		
LEED buildings?		
If yes, what are the LEED rating levels?		
Policy on future LEED buildings?		
Energy Efficiency and Conservation		
Number of window-based air conditioners installed?		
Unoccupied rooms heated or cooled?		
Vending machines lighted? refrigerated? use Vending Misers™?		
Two examples of excessive/extraneous indoor lighting?		
Three examples of unnecessary outdoor lighting?		
Washers and dryers labeled as Energy Star™?		
Alternative Energy		
Energy produced by solar energy?		
Other forms of alternative energy installed?		
RECs purchased?		
Water Conservation		
How many low-flow water sink faucets installed?		
How many waterless or low-flow toilets installed?		
Are water conservation measures used in the landscaping?		
Solid Waste Management		
How many water-bottle-filling stations?		
Do the food kiosks or stores provide single-use plastic bags? Straws? Expanded polystyrene foodware?		
How many recycling stations?		
Pro-environment messaging installed?		
Move-out program for dorms?		
Surplus or reuse center on-site?		

Food		
At a food kiosk or cafeteria, is locally produced food available?		
Is the cafeteria trayless?		
Pro-local food messaging installed?		
Community gardens?		
Food waste composted on-site? off-site?		
Alternative Transportation		
How many bike racks? Are there enough? Are they convenient?		
Bike share program?		
Carpool parking lots or spaces available?		
Car or ride share program?		
Site served by public transportation?		
Charging stations for electric vehicles?		
Others		

Analyzing and Interpreting Environmental Data

OBJECTIVES

- Explain the purpose of a hypothesis.
- Collect basic environmental data.
- Use data analysis software to conduct basic statistical tests for significance.
- Interpret your results.

KEY TERMS AND CONCEPTS

- ✓ Aternate hypothesis (H_A)
- ✓ Data
- ✓ Dependent variable
- ✓ Independent variable
- ✓ Mean
- ✓ Microclimate
- ✓ Null hypothesis (H_O)
- ✓ p-value
- ✓ Spreadsheet
- ✓ Tables
- ✓ t-test

TIME REQUIRED: 60 minutes in class, 30 minutes in computer lab.

Reversing the decades-long trend of movement to the suburbs, Americans are increasingly moving back to urban areas. This dramatic demographic shift has spurred development and redevelopment of urban areas that have also involved the transformation of the natural environment into buildings, parking lots, sidewalks, and roads. This transformation of the natural environment is also occurring in fast-growing metropolitan areas that are sprawling, such as Houston. As an environmental scientist, the logical question is how does this development impact the local urban environment? Building materials such as concrete and asphalt have increased heat-storage capacity, which means that heat collected during the day is slowly radiated back into the environment at night. As shown in Figure 2.1, the phenomenon of higher temperatures in urban areas is called the **Urban Heat Island**, where urbanized areas are significantly warmer than the surrounding rural areas, resulting in microclimates throughout the area. (**Microclimate** is the climate of a small, specific place within an area as contrasted with the climate of the entire area.) To be sure, it is not just the use of materials, but the heat from vehicles, machinery, and heating and cooling systems in addition to localized wind speeds and directions. Increased temperatures can necessitate increased energy demand for cooling, which further increases carbon emissions. Knowledge of the heat absorption properties of building materials is crucial to the sustainable design of communities to reduce the Urban Heat Island effect and to design buildings to reduce cooling and heating costs. The selection of indoor lighting is also important because using energy-efficient lighting means less heat is generated from the lights, which means a reduced need for cooling during warmer months. Again, less energy consumed means fewer carbon emissions.

As shown in Figure 2.2, application of the scientific method involves the formulation and testing of a **hypothesis**, which is a tentative/predictive, testable answer to a specific scientific

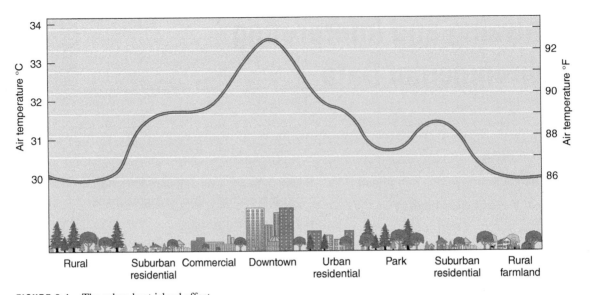

FIGURE 2.1 The urban heat island effect

Source: Raven, P.H., L.R. Berg, and D.M Hassenzahl. 2010. Environment. 7th ed. John Wiley & Sons, New York (p. 271).

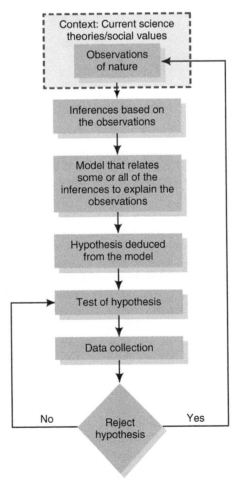

FIGURE 2.2 The scientific method

question generally based on an observation. **Data**, which are values of qualitative or quantitative variables that belong to a set of items, is necessary to test your hypothesis. In this lab, you will be exploring the phenomena of heat generation and absorption by formulating hypotheses, collecting data, inputting data into **spreadsheets**, conducting basic statistical tests, analyzing your results, and, finally, interpreting your results. These steps will help you develop the basic skills necessary for employing the scientific method to examine environmental issues.

- Excel software

- Fluorescent and halogen or incandescent (of the same wattage)

- Infrared (IR) thermometer

- Paper, to cover ~0.5 m^2 of a window

- Sampling grid, 16 squares (50 cm × 50 cm or 1 m × 1 m squares with internal grids) that you can use to mark out 16 sampling areas

1. FORMULATING HYPOTHESES

The first step is the formulation of a series of testable hypotheses. Based on the provided observations, what do you believe the results from the experiment will be? Often, a **null hypothesis** (H_O) is used, which states simply that there will be no effect or change from a particular experiment. In contrast, the **alternate hypothesis** (H_A) predicts that there will be a change or difference or what the difference or change will be.

Observation #1: *Thermal energy loss through windows.* Windows have lower resistance (this is referred to as the *R-value*) to thermal transfer compared to insulated walls. If this is correct, the indoor surface temperature of a window should be different than the surface temperature of an adjacent insulated wall. If it is colder outside, the surface temperature of the window will be colder than the wall. Conversely, if it is warmer outside, then the surface temperature of the window will be warmer than the wall. If there is a temperature difference between the wall and the window, we can conclude that there is a difference in heat transfer through the two surfaces.

Observation #2: *Absorption of thermal energy of different materials.* Asphalt is a common material used for roofs, streets, parking lots, and sidewalks. Given that asphalt is black, one would expect that it would absorb more thermal energy (heat) from the sun compared to nonasphalt materials that are not black. If this is correct, the surface temperature of the asphalt should be higher than that of the nonasphalt materials.

Observation #3: *Impact of shade on thermal absorption.* A noticeable feature in some urbanized areas is the lack of trees, or at least significantly fewer than suburban and rural areas. While there are many benefits to trees, one could ask if they are important regarding the Urban Heat Island effect. Can shade from trees reduce heat from sunlight from being absorbed by building materials such as concrete, brick, asphalt, metal, or wood? If this is true, then one would expect surface temperature of a shaded material to be lower than that of the same material that is not shaded.

Observation #4: *Heat contribution of lighting.* Anything that uses electrical power generates heat. Thus, an object that uses less electricity to perform the same function as a similar object should generate less heat. For example, fluorescent lighting is more efficient than halogen or incandescent lighting; does it also generate less heat?

Below, there are four null hypotheses related to the above observations. For each null hypothesis (H_O), write an alternate hypothesis in which you construct a predictive, testable statement. Be sure there is an **independent variable** (the variable that is changed in a scientific study) and a *dependent variable* (the variable that is measured in a scientific study). Below, the independent variable is underlined and the dependent variable is italicized.

A. Complete the alternate hypothesis (H_A) for observation #1.
H_O: *Heat loss* is the same through a <u>window</u> and an <u>insulated wall</u>.
H_A:

B. Complete the alternate hypothesis (H_A) for observation #2.
H_O: The *surface temperature* of <u>asphalt</u> and <u>nonasphalt</u> (lighter colored material such as concrete, bare ground, grass, etc.) exposed to sunlight is the same.
H_A:

C. Complete the alternate hypothesis (H_A) for observation #3.
H_O: The *surface temperature* of a <u>building material in the direct sun</u> is the same as the *surface temperature* of the <u>same building material in the shade</u>.
H_A:

D. Complete the alternate hypothesis (H_A) for observation #4.

H_O: The *surface temperature* of a <u>fluorescent light</u> is the same as the surface temperature of an <u>incandescent light</u> (or halogen light).
H_A:

Following the instructor's approval of your alternate hypotheses, you are now ready for the next step, collecting data to test your hypotheses.

2. DATA COLLECTION

A. Heat loss experiment

Using the IR (infrared) thermometer, take 16 temperature measurements of a window[1] using a basic grid pattern or using a sample grid (Figure 2.3) (be sure to take a measurement within the middle of each square of the grid). Then take 16 measurements of an adjacent wall also using a grid. Record

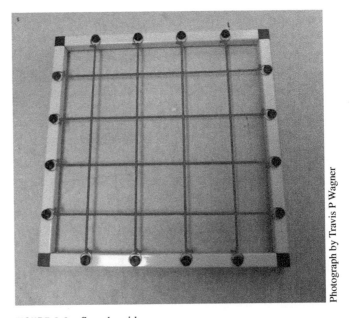

FIGURE 2.3 Sample grid

Photograph by Travis P Wagner

---

[1] To obtain accurate measurements with the IR thermometer, the window must have a paper layer on the window. The paper layer will not affect the temperature.

Table 2.1   Data Collection Table for the Heat Loss Experiment

Sample #	Window (°C)	Wall (°C)
1		
2		
3		
4		
5		
6		
7		
8		
9		
10		
11		
12		
13		
14		
15		
16		

your data in the data collection sheet (Table 2.1), which you will later copy into an Excel spreadsheet. (Or, if you have a laptop or tablet, create and input your data into the Excel spreadsheet directly.)

B. Absorption of thermal energy experiment

Using the IR thermometer and sample grid, take 16 temperature measurements of a designated area of asphalt and 16 measurements of an adjacent surface area that is not asphalt (be sure to note what type of material it was). Record the temperatures into the data collection table (Table 2.2).

Table 2.2   Data Collection Table for the Absorption of Thermal Energy Experiment

Sample #	Asphalt (°C)	Nonasphalt (°C)
1		
2		
3		
4		
5		
6		
7		
8		
9		
10		
11		
12		
13		
14		
15		
16		

Table 2.3 Data Collection Table for Impact of Shade Experiment

Sample #	Shade (°C)	Nonshade (°C)
1		
2		
3		
4		
5		
6		
7		
8		
9		
10		
11		
12		
13		
14		
15		
16		

C. Impact of shade experiment

With the IR thermometer, using the sample grid, take 16 temperature measurements of a designated area of shaded building material (wood, concrete, brick, or steel) and 16 measurements of an adjacent area of the same material that is not shaded. Obtain your recordings and record the temperatures into the data collection table (Table 2.3).

D. Impact of heat contribution from lighting experiment

Using the IR thermometer, take 16 temperature measurements of an operating fluorescent light (be sure to take measurements from all around the lighted surface portion to capture the entire surface area) and 16 measurements of an incandescent light (or halogen) with the same wattage rating as the fluorescent light (be sure to note what type of light it is and the wattage of both). Record the temperatures into the data collection table (Table 2.4).

3. ANALYZING YOUR DATA

How do you know if you can reject the null hypothesis? In other words, can you state with confidence that temperature is lower on the surface of windows, temperature of asphalt is higher, temperature of shaded building material is lower, or that temperature of LED lights is lower? Yes, if you use statistics to analyze the results.

You should be familiar with descriptive statistics, mean, median, and mode, which can help you determine if there is a difference, but, as you will see, a slightly more sophisticated analysis is needed to make a more confident determination.

For *each* of the four experiments, using Excel, you must first transfer your data from your four tables into four separate spreadsheets. In Excel, on the bottom of the spreadsheet, there is a tab labeled Sheet 1. Right-click the tab and select Rename, then rename this sheet, *Heat Loss Window*. Then, click the addition symbol (+) to create a new separate spreadsheet within this file. For each sheet, use the same format and titles as your initial data collection sheets for the columns and rows (e.g., Heat Loss Window, Asphalt Temperature, Shade Temperature, and Light

Table 2.4 Data Collection Table for the Heat Contribution of Lighting Experiment

Sample #	Fluorescent (°C)	Incandescent (°C)
1		
2		
3		
4		
5		
6		
7		
8		
9		
10		
11		
12		
13		
14		
15		
16		

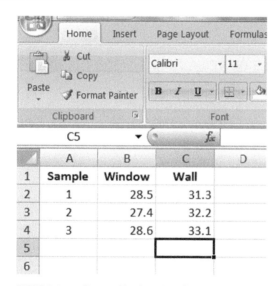

FIGURE 2.4 Data collection sheet in Excel

Temperature) and then input your data from the appropriate data collection table. It should look like Figure 2.4. (This is a Windows version; Mac version will be different.)

4. CALCULATING THE MEANS

For *each* experiment, calculate the **mean** for each column. For each column, put your cursor in the empty cell immediately below the last entry. In the tool bar, select the insert function icon (see Figure 2.4) and then search and select the function, AVERAGE. Make sure that the appropriate column and row numbers are included within your selection (e.g., B2:B16), and then press Enter. Do this for each column and for the other experiments in the other sheets. Label this row in your spreadsheet as Mean.

5. CONDUCTING *t*-TESTS USING EXCEL

Now you will be conducting a more sophisticated statistical analysis. Why? Statistics uses mathematics to quantify uncertainty that is inherent in the scientific method. That is to say, it provides an objective tool to help scientists answer a research question (in this case to test a hypothesis) with greater confidence. In the previous section, you calculated the mean which is a type of descriptive statistics (*descriptive* means it is a summary or description of the data). In science, we need to draw inferences about a process or population under study using *inferential statistics*. Inferential statistics helps scientists identify whether the results are random or not; that is, is there really a difference or impact or is it merely a result of randomness? In this task, you will be conducting *t*-tests, which are very simple yet powerful statistical tests to compare two means. This comparison will enable you to answer the question: Is A significantly different than B?

Using your Excel spreadsheet, you will be conducting two different types of *t*-tests (the standard *t*-test for the window [type 1] and a paired *t*-test [type 2] for the other three experiments).

For *each* experiment, calculate the *t*-test for all the data (you will have one **p-value** for *each* experiment). Copy and paste your row of means down a row to make room for the *p*-values directly under the columns of data. Put your cursor in the empty cell immediately following any last entry other than the sample number. In the tool bar, select the insert function icon then search and select the function, TTEST (you may first have to go to Category and select "statistical"). As shown in Figure 2.4, in Array1, highlight all the entries in the first data column and then hit Enter (it should read B2:B16). Repeat this for Array2 and the second data column (it should read C2:C16). For Tails, select 2. For Type: for the Heat Loss, Asphalt, and Shade experiments, enter 1 and then select OK, and for the Lighting experiment, enter 2, and then select OK. The result will be the *p*-value. Label this row of data "*p*-value."

6. CREATING A TABLE OF RESULTS IN A WORD DOCUMENT

For *each* experiment, you will be creating a **table** (a *table* is a means of arranging data in rows and columns). Format the table with lines, headings, number format (only use two decimal places for the mean—for example, 23.45—as Excel calculates a level of precision that is far excessive for these experiments) as shown in Table 2.5. This table will be the results in your lab report as described below.

Table 2.5 Example of the Results Table

Experiment	Mean Temp (°C)	*p*-Value
Heat loss		
Window		
Wall		
Absorption of thermal energy		
Asphalt		
Nonasphalt		
Impact of shade		
Shaded material		
Nonshaded material		
Heat contribution from lighting		
Fluorescent light		
Incandescent light		

7. INTERPRETATION OF THE RESULTS

Now that you have collected your data and analyzed your results, you must explain what it all means. Specifically, you are trying to determine if the results allow you to reject the four null hypotheses or not. How do you do this? Your calculations of the p-values hold the answers. If the p-value is below 0.05, this is generally considered statistically significant, while a p-value of 0.05 or greater indicates no significant difference between two groups. Another way of looking at this is to think of the p-value as a percentage and if the p-value is less than 0.05, you can state that you are 95% confident that there is a significant difference between the two groups (in science, 95% confidence level is a common minimum for statistical differences) and thus you can reject your null hypothesis, which stated that there was no difference. At this point, you can only state that there is a significant difference and you may accept your alternate hypothesis if it predicts a significant difference. You must use the means that you calculated earlier to determine which value is higher or lower.

For each null hypothesis, state whether you reject or accept the hypothesis referencing your p-value. Then, using the mean, if you can, state which one is higher and which one is lower.

It is now time to prepare your written lab report, which is patterned after a standard scientific paper. Be sure to use the following headings and associated contents.

WRITE-UP

Title	A short, concise title that indicates the subject matter of this lab.
I. Introduction	A brief paragraph about the purpose of this lab.
II. Background	A brief discussion of the relevancy of this lab activity. For example, discuss the environmental implications of the Urban Heat Island effect, energy demands of building materials, or the role of urban energy demand and global climate change. In this section, it is imperative to cite references. Finally, this section ends with each of your stated hypotheses.
III. Methods	For each experiment, describe how (and with what instruments and materials) you collected the data.
IV. Results	What were your results? Insert the table of the results as discussed above. Use graphs if they add value to your presentation.
V. Discussion	Explain if you rejected or accepted each hypothesis and discuss what you found relating to your background above.
VI. References	An alphabetized list of all cited references in proper format.

Environmental Science in the Media

OBJECTIVES

- Identify the major challenges of accurately communicating science in the media.

- Understand the difference between an environmental condition and an environmental problem.

- Describe, critique, and present a major environmental science controversy reported in the mass or social media.

KEY CONCEPTS AND TERMS

✓ Causes

✓ Information vacuum

✓ Journalistic norm of balance

✓ Newsworthiness

✓ Peer review

✓ Symptoms

TIME REQUIRED: ½ hour in class, 1 hour in the library or computer lab, and ½ hour to present the results to the class.

INTRODUCTION

Historically, for most people, the traditional news media (e.g., newspapers, television, and radio) has been the primary source of news related to science and environmental issues. This, however, is changing. According to the Pew Research Center (Mitchell, Gottfried, Barthel, and Shearer, 2016), adults in the United States still get their news from television (57%) while online sources are rapidly increasing and are now used by 38% of adults. This is followed by radio (25%) and traditional newsprint (20%). Facebook and Twitter are increasingly becoming popular sources for news headlines, which include environmental issues.

So, what makes news? First, it is crucial to understand that news is one mechanism to attract readers, which supports the media source through subscription or advertising. News stories are those defined as newsworthy, which is not the same as important. There are 10 primary factors of **newsworthiness**. Moreover, reporters and editors (known as communication gatekeepers) generally are not trained scientists, but journalists, which can affect their ability to distill and report science accurately. Environmental issues tend to be emotionally charged because they can adversely impact humans, animal health and welfare, a social or racial minority and are likely to have significant economic impacts. Controversy (aka, conflict) makes reading interesting—it may be profitable to also inflate or focus on the extremes as a way to charge an issue and make controversies seem greater than they are, which attracts readers and helps promote sharing and retweeting.

Another important aspect of journalism relevant to reporting on environmental science is the journalistic norm of "balance." With this norm, journalists give both sides of an issue equal weight and voice even if they do not have equal evidence of the issue. This balance approach is often cited as a reason why there is a large percentage of the public who falsely believe that there is considerable disagreement in the scientific community over the role of humans in global climate change. Consumers of news need to be critical thinkers.

As an environmental scientist or decision-maker—and we are all decision-makers when we go to the polls to vote—the fundamental question is whether something is defined as an environmental condition versus an environmental problem. Environmental problems are issues that warrant public action to address or solve because of significant adverse impacts to human health, the environment, or the economy. In contrast, an environmental condition is not defined as a problem, that could be categorized as a concern or nuisance but does not yet warrant public action to address. Key to this definition is the number, type, and impact of the symptoms and the causes of the problem. **Symptoms** are the empirical evidence that tell us if it is an environmental problem such as the number of people, animals or habitat harmed, or the economic impact to society. **Causes** of the problem are those factors that are actually creating the problem. For example, a fish kill is a problem. We know that because there are dead fish and people are complaining of the smell (the symptoms). But we need to know what caused the fish kill, which may be reduced dissolved oxygen or a chemical spill.

A challenge in trying to determine the difference between a condition and a problem is the role of newsworthiness, which can skew the perceived impacts of a potential problem. Thus, the use of inflammatory and exaggerative language can alter one's opinion or perception even if the facts do not exist—scientists prefer facts as opposed to claims. Herein lies another dilemma: if an issue becomes known, we need to look to facts and knowledge. However, if it is a new issue that receives media attention, an **information vacuum** can exist, which is a circumstance where there is insufficient factual information to assess an issue and/or suggest solutions. If an information vacuum exists, the supply of opinions, stories, and evidence (factual, anecdotal, or false) quickly fills the vacuum because of the demand for information. If the information vacuum has been filled with false or wrong data, it still becomes the dominant knowledge base for the issue. Science, however, is designed to be slow because the goal is accuracy, not expediency. This means that by the time the issue has been studied by science and **peer-reviewed** articles have been published, there may already be entrenched perspectives of the issue that may conflict with the science.

The purpose of this lab is to evaluate the role that the media plays in presenting environmental information. Specifically, we will focus on the positive and negative contributions of media. Communicating science is not limited to writing; scientists are expected to present information orally to the public, to peers, to the media, and to decision-makers. Hence, you will be making a presentation to the class in addition to a written report. Because of the increasing presence and need to communicate science in a public forum, for this lab, the classroom is intended to replicate your public forum.

THE 10 FACTORS OF NEWSWORTHINESS

1. *Bizarreness/oddity:* Is it a weird or unusual situation?

2. *Conflict:* Is there significant controversy, argument, or disagreement?

3. *Human interest:* Does it involve a named person or family?

4. *Impact:* Are many people or organisms impacted?

5. *Prevalence:* Has it already been in the news recently?

6. *Prominence:* Does it involve famous people, landmarks, or corporations?

7. *Proximity:* Did the issue happen close to home?

8. *Timeliness:* Is it brand new, is it an exclusive?

9. *Scandal:* Scandalous action or behavior

10. *Extremes:* The worse, the least, the biggest, or the newest

---

No special materials are needed for this lab other than access to the Internet.

MATERIALS

TASKS

1. As a class, generate a list of current and recent environmental controversies. Make sure they are sufficiently narrow; global climate change is way too broad, but the role of cows in climate change may not be. As a group (preferably of three), you will select one of the controversies and then divide up the major media sources: one person will search Facebook, one person will search Twitter, and one person will search online newspapers.

2. Using the list of questions below, evaluate the positive and negative contributions of the media coverage. Focus on the media's ability to impart the information necessary for the

public and decision-makers to make properly informed decisions on environmental issues. You will be examining the factors of newsworthiness to assess if the issue is just a condition or a problem that warrants pubic action. You also need to look for inflammatory language and whether the **journalistic norm of balance** was used.

3. As a group, you will be presenting your results to the class by comparing and contrasting the treatment of the issue by the different media sources. (An option is to submit a typed report instead of the class presentation.)

QUESTIONS

1. What is your topic?

2. What was your source (Facebook, Twitter, name of the online newspapers, etc.)?

3. Select four different articles, posts, or threads. What are the titles/headlines?

4. What are the reported symptoms of the problem?

5. What are the reported causes of the problem?

6. To what degree was the journalistic norm of balance used? (That is, did the author cite experts representing both sides, pros and cons, of the issue?). If so, what are the qualifications/expertise of the expert opinions/sources cited or quoted (do this for each person or organization)?

7. Summarize and discuss the manner and degree to which vested interests (e.g., economics, jobs, health) or emotional interests are discussed in each article.

8. Which of the factors of newsworthiness likely were used to select the story?

9. To what degree was exaggerative or inflammatory language used to describe the issue or to describe the impacts of the issue?

10. Would you define the issue as an environmental condition or an environmental problem and why?

CLASS PRESENTATION

The following below are some for this lab, as a group, you will be presenting your findings to the class using the following suggested headings as title slides. The point of a group presentation is to compare and contrast the same issue as viewed through different media sources. Thus, each person will address each of the headings below specific to their media source. One approach is to include a table for each slide that allows for the comparison of the three media sources/news sources:

Title: (name of the issue)

   I. **Media Sources** (what media source did you search, when, how, etc.?)
  II. **Symptoms and Causes of the Issue**
 III. **Newsworthiness**
 IV. **Journalistic Norm of Balance**
  V. **Condition or Problem?**

REFERENCES

Mitchell, A., J. Gottfried, M. Barthel, and E. Shearer. 2016. 1. Pathways to News. Pew Research Centers Journalism Project RSS. July 7.

Sustainability and Business

OBJECTIVES

- Research and report the challenges for a business to create, adopt, and maintain sustainability efforts.

- Describe the manner in which a sustainable business balances people, profits, and the environment.

KEY CONCEPTS AND TERMS

✓ Cradle-to-grave

✓ Energy Star

✓ EPEAT = Electronic Products Environmental Assessment Tool

✓ Greenwashing

✓ LEED = Leadership in Energy and Environmental Design

✓ Life-cycle analysis

TIME REQUIRED: For virtual field trips, 1 hour plus additional time to contact the corporate sustainability representative; for site visit, 1 hour, arranged in advance.

INTRODUCTION

"Going green" and being sustainable have become popular as the corporate world responds to environmental and social concerns, especially in relation to global climate change and social justice. Becoming "green" does not mean just making "green" products and it definitely does not mean greenwashing. (**Greenwashing** is when a corporation or organization deceptively uses the environment, including concepts, symbols, and slogans, to sell products or to sell a corporate or organizational persona when the product, corporation, or organization would not normally be viewed as "green" or sustainable.) Being sustainable is a broad concept that is not limited to environmental concerns. As shown in Figure 4.1, the concept of sustainability rests on three pillars: environmental, social, and economic. Thus, businesses attaching a label or claim of sustainability are also expected to offer living wages, engage in fair trade and fair labor practices yet still make a profit. While most companies may not be able to claim to be a fully sustainable company, they can adopt components or principles of sustainability. While large corporations may have a full-time sustainability office or officer, small- and medium-sized companies do not. However, many third-party organizations provide assistance or certifications to help companies adopt sustainable practices. For example, a company can adopt specific aspects of sustainability, which include purchasing only **Energy-Star**-rated appliances and office equipment; computers that are **EPEAT** registered under the Green Electronics Council; or constructing buildings that are **LEED** certified by the U.S. Green Building Council.

Being sustainable can also be profitable. Examples include Ben & Jerry's Ice Cream, which became famous in the 1980s and continues to thrive despite having to make some concessions to the larger corporate world. Tom's of Maine has become known as a proponent of natural health-care products and ethical corporate behavior. Other well-known companies that have adopted various aspects of sustainability include Starbucks, Burt's Bees, Adobe

THIRD-PARTY ORGANIZATIONS AND CERTIFICATIONS FOR SUSTAINABLE BUSINESS PRACTICES

- EPEAT
- Fair Trade USA Certified
- Forest Stewardship Council
- Green America
- Green Business Bureau
- Green C Certification
- Green Plus
- GRESBE
- LEED
- ParkSmart

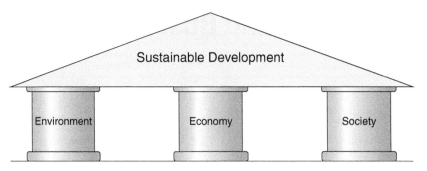

FIGURE 4.1 The three pillars of sustainable development

software, and Whole Foods. It is important to understand that no company is perfect; they must balance the profit expectation while becoming more socially, economically, and environmentally sustainable. Being sustainable can be a complicated undertaking.

This lab focuses specifically on the environment pillar of sustainability, which includes a variety of actions such as reducing the environmental impacts of manufacturing and packaging products as well as designing products so that they can be reused, recovered, or recycled at their end of life. The holistic approach to identifying and reducing impacts of a product from its **cradle to grave** (or cradle-to-cradle) is the foundation of the **life-cycle analysis** strategy: examine impacts at each step and then identify appropriate actions to reduce the generation of pollution or waste or reduce the consumption of energy. "Cradle" refers to the very start where raw materials are obtained to create the product and "grave" refers to the ultimate environmental fate after the product has ended its useful life.

In this lab, you will be identifying an example of a sustainable company (or a company adopting sustainable practices) and then through a site visit or virtual field trip, you will be preparing and submitting a technical report on the company's sustainability status. As you complete your investigation, keep these questions in mind: What does it take to be a socially and environmentally responsible company? How do we recognize these companies? If you conduct interviews or use reference materials, be sure to cite them in a reference section.

| TASKS | The following questions should be used as a guide for the virtual or actual field trip and to prepare your write-up in the format of a technical report: |

1. Why did you select this company? What are its characteristics (what do they produce, how long has it been in business, how many employees, and where is its headquarters)?

2. Locate the company's mission or policy statement related to the environment, sustainability, or global climate change. How well does it incorporate each of the three pillars of sustainability?

3. Go to the company's website; does it advertise itself as green or sustainable? If so, how? List all the specific statements and/or claims that the company makes regarding sustainability. In addition, answer the following questions:
   A. Is the company ISO 14001 certified? Describe what is covered under ISO 14001.
   B. Does the company have any LEED-certified buildings?
   C. Has the company adopted a policy regarding Energy Star?
   D. Has the company adopted a policy regarding EPEAT?
   E. Has the company adopted any other type of green/sustainability policy or practice or has it obtained relevant certification? Consult the box "Third-Party Organizations and Certifications for Sustainable Business Practices" on various sustainability-related programs and third-party certifications to determine if your company has obtained any of the certifications or others.

4. Based on the type of company, what environmental factors are involved in the day-to-day operations of the facility (e.g., water discharges, air emissions, waste generation)?

5. If the company manufactures products, what type of raw materials does it consume and where do they come from?

6. What happens to the company's product at the end of the product's useful life? What happens to the packaging used to contain and/or ship the product?

7. Is there any data (and if so, what does it say) provided by the company on environmental impacts including air emissions, water discharges, spills, hazardous waste, and solid waste. What do the data indicate?

8. Using the U.S. Environmental Protection Agency's Enforcement and Compliance History Online (Echo) website (https://echo.epa.gov) search for your company to assess their compliance with environmental regulations. What were the results of your search?

9. Is there any data (and if so what does it say) provided by the company on energy consumption, energy conservation or efficiency, or alternative energy? What do the data indicate? How does a company like this incorporate a green/sustainable philosophy while being profitable (be specific)?

10. Discuss two specific lessons you learned about the company regarding the adoption of sustainability in the for-profit world?

WRITE-UP

The write-up for this lab is a typed, technical report. Based on the subject of the report, diagrams or photographs (provided the company gives permission) would be helpful. The following are the suggested headings for your technical report:

Title: (name of the company, product, or organization)

  I. Introduction (why, what, where, and who)

 II. Sustainable Practices (describe processes, programs, and/or efforts related to sustainability; include data and diagrams)

III. Sustainability Certifications (identify any certificates related to the three pillars of sustainability)

IV. Challenges in Becoming Sustainable (discuss the challenges, barriers, and limitations facing the company in becoming more sustainable)

 V. Next Steps (discuss the next steps for the company or offer recommendations on how the company can improve its sustainability)

VI. References Cited

Environmental Site Inspection

OBJECTIVES

- Learn how to conduct visual surveys to inspect buildings, structures, and grounds to determine if there are indicators of health or environmental risk, harm, or damage.

- Complete an environmental inspection form similar to those used in professional Phase I site assessments to prepare a technical report.

KEY TERMS AND CONCEPTS

✓ Environmental Site Assessment

✓ Phase I Site Assessment

✓ Level I Survey

TIME REQUIRED: 30 minutes to 1 hour for computer research, at least 1 hour in the field doing the site inspection.

INTRODUCTION

Many concepts that you learn in environmental science involve the prediction or evaluation of environmental impacts. In this lab, you will go to the field and identify actual health and environmental impacts by conducting a site inspection using a standard site inspection form, which will form the basis of a technical report.

Site inspection forms are used by environmental professionals to assess the potential for contamination and/or negative environmental conditions primarily for clients interested in purchasing property (e.g., ASTM, 2016). That is, because of potential liability concerns, prospective purchasers of property want to know if there are potential or actual environmental problems. This is done through a **Phase I Site Assessment**, which is also known as an **Environmental Site Assessment** or **Level I Survey**. Phase I assessment includes a review of historical records, federal, state, and local files, and a visual inspection to look for actual or potential environmental problems (sometimes referred to as "recognized environmental conditions" or RECs). Many consulting companies and agencies have their own forms for these assessments.

For this lab, you will conduct background research on your site, inspect the site and complete an inspection form, and summarize the procedures and findings in a technical report.

TASKS

Typically, you will conduct some preliminary research on the property before the actual inspection. This research can readily be done using Google Earth, and was traditionally done by hand with aerial photographs, historic maps such as the Sanborn Fire Insurance maps, USGS topographical maps, National Wetlands Inventory maps, and property deeds of adjacent landowners. For this lab, we will not be interviewing former owners and/or workers on the site, but you will need to review government records pertaining to the site, such as those provided by the U.S. Environmental Protection Agency's "My Environment" to identify what is in your zip code region in terms of environmental conditions, including Superfund sites,[1] permitted water discharges, permitted air

---

[1] **Superfund,** officially known as the Comprehensive Environmental Response, Compensation, and Liability Act (CERCLA), covers former or current hazardous waste sites that are abandoned or where the owner is no longer able to pay for cleanup. Superfund establishes strict liability (i.e., there is no need to show negligence) for property owners where hazardous substances present a threat to human health or the environment, regardless of when the hazardous substances were placed there.

discharges, hazardous waste generators or management facilities, and facilities reporting toxic releases. A local search (municipal records offices) may also be done for the property of interest and for adjacent and any nearby properties that could affect your property of interest (e.g., properties that are in the direction where the wind primarily comes from upwind or upslope).

1. BACKGROUND RESEARCH

To conduct the background research for this lab, use the My Environment tool (https://www3. epa.gov/enviro/myenviro) to identify on-site and adjacent areas of potential concern.

A. Input the site's zip code.

B. Check all boxes of interest under MyMap features. To learn more about the items, click them for additional information.

C. The map will show any emergency incidents, Superfund sites, hazardous wastes, toxic releases, brownfields, and hazardous waste sites. Click "Redraw map."

D. Under MyAir and MyWater features, also on the Web page, check to see what information is available near your site.

E. What facilities or issues or reported events (spills, discharges, exceedances) did you find within 1 mile[2] of the site?

F. For each facility, list the regulated activity (air emissions, hazardous waste, water discharges, etc.).

G. Is the facility or event of potential concern to your property? Why?

2. SITE INSPECTION

Go to the site designated by your instructor and fill out the Inspection Form and Inspection Checklist located below. If the site is a university building, use the approximate environs as the parcel, not the entire campus. You will be looking for evidence of disturbances such as discarded material, discolored or dead vegetation, stained soil, objects protruding from the ground, mounds or depressions, and odd smells. Do not leave blanks or skip any items on the form. If the item is not applicable, mark it "NA"; if an item was not inspected, mark it "NI" and be sure that the reader knows why it was omitted. For any item checked "yes," be sure to write a comment and take a photograph to help illustrate your comments.

The write-up for this lab is a technical report as you need to communicate your findings to your client, in this case, the lab instructor. This will require you to prepare your technical report with a copy of your inspection checklist as an attachment. Use the following format in your report. (And do not forget the completed inspection form.) The tone of your report should be neutral and fact-based.

WRITE-UP

Title: (name of issue)

  I. Introduction (what you did, when you did it, why you did it, and where you did it)

 II. Background Research (results of your background research)

III. Site Inspection (results of your site inspection including photographs)

---

[2] In the United States this is an exception to the normal rule of using the metric system rather than English because the general public relies on the inspection information.

IV. **Recommendations** (e.g., "No actual or potential problems" or "Further investigation is needed because . . .")

V. **Completed Inspection Form**

REFERENCES

American Society for Testing and Materials (ASTM). 2016. E1527-13. Standard Practice for Environmental Site Assessments: Phase I Environmental Site Assessment Process. West Conshohocken, PA. Available at https://www.astm.org/Standards/E1527.htm (verified 23 June 2017).

Environmental Protection Agency (EPA). My Environment [Online]. Available at https://www3.epa.gov/enviro/myenviro (verified 23 June 2017).

Google. Google Earth [Online]. Available at http://earth.google.com (verified 23 June 2017).

ENVIRONMENTAL INSPECTION FORM

Date of inspection _____ Inspected by _____

Site conditions (anything that affects your viewing, like rain, leaves, covered in snow, other visibility factors)

Street and mailing address for the inspected property: _____

_____ Include grid coordinates if available _____

General description (e.g., ¼-acre residential lot, manufacturer, government, school, commercial business, etc.) _____

Client: (in this case, it might be your instructor) _____

Current Use of Property: Residential _____, Commercial _____, Industrial _____, Government Institution, Agricultural _____, Forestry _____, Other _____

Amount of Raw (Undeveloped) Land: _____

Past Use of Property: Residential _____, Commercial _____ Industrial _____, Agricultural _____, Forestry _____, Other _____

General Field Observations

> **WARNING:** The purpose of this lab is educational. Your health and safety are of paramount importance. Do not enter private property. Do your inspections in teams of two or more, if possible. Wear safety vests if near any roads. Stay outside of structures. Do not, under any circumstance, open, touch, move, or otherwise manipulate any unknown containers, tanks, and so forth. If you observe such items, mark it as such on your inspection form. If you have a concern, bring it to the attention of the instructor. *Avoid contact!*
>
> After reading the above statement, sign below that you understand the importance of health and safety for this lab.
>
> Signature _____ Date_____

OUTSIDE INSPECTION (Comment on all YES items)			
Underground storage tanks (USTs)	YES	NO	UNK
Known or observable underground storage tanks? How many? _____			
Any evidence of underground storage tanks?			
▪ Fuel dispensers			
▪ Vent tubes			
▪ Hoses			
▪ Manhole covers			
Any evidence of soil staining?			
Comments			
Aboveground storage tanks (ASTs)	YES	NO	UNK
Known or observable aboveground storage tanks? How many? _____			
Any evidence of aboveground storage tanks?			
▪ Signage			
▪ Berms			

	YES	NO	UNK
▪ Pipes			
▪ Heating oil			
Any evidence of soil staining?			
Comments			
Drums/containers	YES	NO	UNK
Known or observable drums/containers?			
How many? _____			
Any evidence of soil staining?			
Comments			
Ecological conditions	YES	NO	UNK
Wetlands (on-site or adjacent)?			
How large? _____			
Floodplains (on-site or adjacent)?			
Unique or critical habitat (on-site or adjacent)?			
"Threatened" or "endangered" species (on-site or adjacent)?			
Brooks/streams/rivers (on-site or adjacent)?			
Ponds/lakes (on-site or adjacent)?			
Earth disturbance (excavation, craters, subsidence, or cracking)			
Vegetation stains, matting, discoloration			
Comments			
Miscellaneous conditions	YES	NO	UNK
Electrical transmission lines present?			
Discarded tires			
On-site sewage disposal (e.g., septic tank)?			
Abandoned drinking water or injection wells			
Mounds or other potential signs of past disposal			
Comments			

For the purposes of this report, do not conduct inside inspection unless authorized by lab instructor. Answer the items below on the basis of exterior conditions.

Lead	YES	NO	UNK
House built before 1978?			
Any evidence of lead-based paint?			
If yes, is paint chipping, cracking, or peeling?			
Evidence of lead piping for water?			
Comments			
Asbestos	YES	NO	UNK
Evidence of asbestos (insulation, pipe insulation, tiles, or siding)?			
Comments			
Hazardous materials	YES	NO	UNK
Pesticides (herbicides, insecticides, etc.)			
Paint, oil based			
Fuels			
Unknown/unidentifiable			
Comments			

Urban Ecosystems

OBJECTIVES

- Conduct an urban inventory through the framework of urban ecology.

- Describe an urban ecosystem and identify its major components.

- Identify and label energy pathways in a particular streetscape setting within an urban ecosystem.

- Compare an urban ecosystem with a more traditional "natural" ecosystem.

KEY CONCEPTS AND TERMS

✓ Block-face

✓ Ecosystem

✓ Energy pathways

✓ Environmental planning

✓ Urban ecology

✓ Urban metabolism

✓ Urban planning

TIME REQUIRED: 2 hours in the field, 1 hour for the write-up.

INTRODUCTION

According to the 2010 U.S. Census, 80.7% of the U.S. population lives in urban areas. Urban areas are also centers of employment. Thus, the immediate environment for most people is an urban area. Urban areas are a type of **ecosystem**, and like any other ecosystem, the urban ecosystem contains units and pathways for the exchange of energy and information (see Figure 6.1). Urban planners and policy makers have recognized the power of incorporating ecosystem concepts into the design and management of urban areas. Periodicals such as *Urban Ecosystems* and *Journal of Urban Planning and Development* reflect research and professional practice involving an ecological approach (urban ecology) to the study of cities. Urban ecology is a common subject in the curriculum of urban planners, designers, and other environmental professions (e.g., Adler and Tanner, 2013). (**Urban ecology** is the study of the relationship of living organisms with each other and with their surroundings in an urban environment.) The description and analysis of flows of the materials and energy within cities is referred to as **urban metabolism**, which was posited over 50 years ago and remains a tool in urban ecology to inventory and analyze technical and socioeconomic processes (Kennedy, Pincetl, and Bunje, 2011). Traditional biological and other environmentally related sciences also consider the effects of urban environments; the effects of urban settlements and the industrial revolution can be felt everywhere, including remote Antarctica.

In this lab activity, you will inventory and analyze a block in an urban area through the framework of urban ecology, which is an approach that replicates what planners, landscape architects, civil engineers, and other professionals do for urban revitalization or related planning or economic development efforts. Collecting urban environmental data for planning and design is complex and often more extensive than as presented in this lab. This lab, however, is designed so that you will have a good idea of the basic data considered in an ecological approach to urban design and management.

In the interest of time, we will be conducting a condensed version of different activities: First, you will sketch a profile of an urban unit, the **block-face** (defined as one side of a street between two intersecting streets or other geographic boundaries). Second, you will inventory the

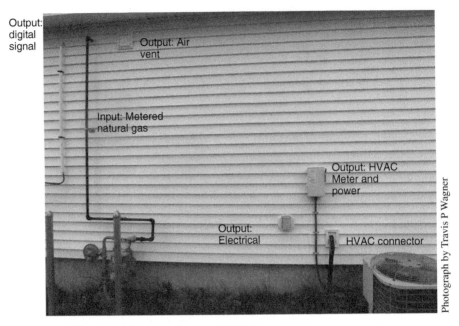

FIGURE 6.1 Example of inputs and outputs labeled on a structure

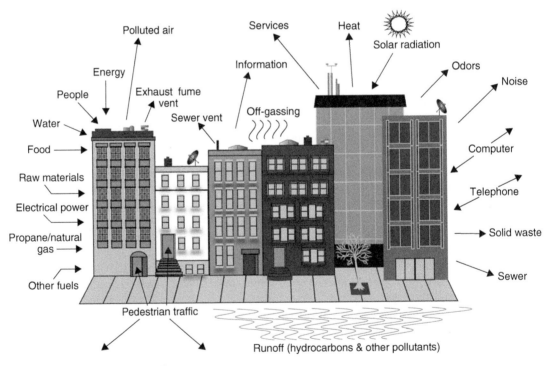

FIGURE 6.2 Urban ecosystem diagram

An urban system's exchange of energy and materials as inputs and outputs. An urban ecosystem contains at least an order of magnitude more energy per unit area than other ecosystems.

different types of **energy pathways** (infrastructure) entering and leaving the unit—driveways, sidewalks, doors, telephones, mail slots, power lines, smokestacks, exhaust vents, any connections to the air, land, water, or other medium (see Figure 6.2 for examples). Third, you will estimate the units or range of units used to measure the use of these pathways (e.g., number of cars using the driveway, amount of smoke going out of the chimney). Finally, you will compare this urban ecology approach to other more "natural" areas that we more commonly think of in ecology, such as a wetland, desert, meadow, or forest.

To approach this lab, imagine that your company or agency has been hired to apply for an urban renewal grant and you want to understand conditions by conducting a preliminary, nonintrusive walkover with a basic inventory. In such a process, one might use transects on a grid or assess socioeconomic or economic spatial sectors.

MATERIALS

- Decibel meter (optional)

- Graph paper, 8.5 × 11

- Handheld counter (optional)

- Safety vest and helmet

TASKS

In teams, you will be assigned a block to profile and inventory (a block-face) (see Figure 6.2, the Urban Ecosystem). The view you have of this block-face is called a "streetscape." In Figure 6.2, the **streetscape** depicts a schematic of an input/output system.

1. At your study site, sketch a profile of the streetscape (also called a "street elevation") in your notebook or a sheet of graph paper. A sketch or photograph is adequate; the important things are the basic information and the concepts. (Convert your sketch or photograph into a CAD drawing if it helps illustrate your streetscape.) Include buildings, fences, roads, driveways (their entrance to the street is called a "curb-cut"), signs, dumpsters, and anything else in the built environment, as well as grass, trees, and other natural features. The streetscape should include the ground and what is likely to be immediately below the ground (i.e., water, sewer, storm water, electrical, telephone, and natural gas). Include the roadways on either side of the block—this is where your study site may overlap with that of another team. If you are not sure whether to include something, put it in; it is much easier to take it out later compared to trying to remember what you saw. Safety first—work in pairs if possible and watch for traffic (your instructor may require you to wear a safety vest and helmet). Stay on public property.

2. Describe your block setting so that someone from another area can find the town, the street location, and the block-face you studied. For example, "We examined the block in Gorham, Maine beginning on the east side of School Street (Route 114) at the intersection of South Street and Maine Street (Route 25) and extending north toward College Avenue. This block-face contains two former churches, a restaurant, a parking lot, and a commercial building with businesses on the first floor and apartments above."

3. List major system inputs and outputs to the structures you illustrated in Task 1, above. Put them on your figure as we have done in the examples (Figure 6.1 and 6.2). In the next step, we will be taking a closer look at the structures as individual components of this urban ecosystem.

4. Inventory the structures you see in your block-face and record the results into Table 6.1. A structure can be a building, a deck, a parking lot, or other built aspect of the environment. Label each structure in a drawing and refer to the label in your table. For example, "Apartment building at 99 Oak Street" or "parking lot for Pancake Restaurant." Label the structure based on what you think its apparent urban function is (e.g., residential. commercial, educational, religious, and administrative). What do you think are the main materials used to construct it, based on the exterior? What is your best estimate on how many people work or live in the structure? How old do you think it is? Does it show any

sign of processes that modified it? Can you tell if it has gone through any changes such as a new addition, a removed portion, a renovation? This is just a "best guess" activity, but urban planners might use it as a basis to go back and do specific research on each structure.

5. Identify the air inputs and outputs for your structures and record these data into Table 6.2. You should have these as general systems inputs and outputs on your Figure 6.1, but here you are looking for specific sources such as chimneys and pipes.

6. Identify the energy inputs and outputs and record into Table 6.3. You are looking for specific sources that can be identified on the exterior of your structures.

7. Identify the solid waste outputs for your structures and record into Table 6.4. What clues can you find for solid waste services, such as waste receptacles, storage areas, and access doors?

8. What water inputs and outputs are there? Describe each input and output and record into Table 6.5. To estimate the amounts of sewer and water, the standard approach is to use billing records or read meters. But since this is a nonintrusive preliminary assessment, you can just identify sewer, water, and other utility pipes if they are visible. If you cannot see them, then

Table 6.1   Structure Inventory

Structure name and label	Type of structure (residential, commercial, industrial, public)	Primary building materials	Estimated number of occupants	Estimated age of structure	Evidence of modifications or changes to structure

Table 6.2   Air Inputs and Outputs

Type	Source(s)
Carbon monoxide	
Sulfur oxides	
Nitrogen oxides	
Hydrocarbons	
Particulates	
Odors	
Electromagnetic	
Other	

Table 6.3   Energy Inputs and Outputs

Type	Source(s)
Electronic (data cables)	
Solid fuels	
Liquid fuels	
Solar	
Odors	
Electromagnetic	

you can still make a good guess if they are there or not. Use the comment section of your table for anything you might want to add to augment your description.

9. Determine the total number of motorized vehicles on your street for a 1-hour period and input into Table 6.6. Do this by counting the vehicles driving past for a half hour period. Multiply the number by 2 to convert the figure into an estimate of what you might expect if you counted for an hour. (If you have multiple lanes of heavy traffic, you might just count the vehicles in one direction for half an hour and multiply by 4 to estimate an hour's traffic in both lanes.) Be sure to record the exact time of the day you did your count. If this was not during the peak travel times (morning or evening), estimate the likely count of vehicles during the peak travel hours. Comment on the capacity of this urban area to accommodate this traffic.

10. Vehicles may enter your site (including parking in front) in generally one of two manners—by design ("destination" trips) or by "capture" (impulse, wrong turn, or advertising that caught the driver's eye, or something similar). The same is true for pedestrians. They may enter a building, a yard, or a driveway. Tally vehicles and pedestrians in the table below for 30 minutes (or for time specified by your instructor). Record these data into Table 6.7.

Table 6.4 Solid Waste Outputs

Type	Source(s)
Construction material	
Manufacturing waste	
Household	
Food industry	
Commercial	
Odors	
Electromagnetic	
Other	

Table 6.5 Water Inputs and Outputs

Type	Description	Location	Comment
Building-mounted drainpipe/gutter			
Storm grate (in road or sidewalk)			
Drinking water pipe			
Fire suppression water pipe			
Sewer pipe			
Sewer grate (road or sidewalk)			
Other			

Table 6.6 Vehicles and Pedestrians Passing on the Street in front of the Block-Face

Motor vehicles	Half hour starting at	Full hour
Pedestrians		
Nonmotor vehicles (e.g., bicycle)		

Table 6.7 Vehicles and Pedestrians Entering or Leaving between Given Time and Day

Motor vehicles	Enter	Exit
Pedestrians	**Enter**	**Exit**
Nonmotor vehicles (e.g., bicycle)		

Table 6.8 Noise at: _____

Time	Structure (location)	Decibel reading	Source(s)

11. Based on your traffic count, estimate the amount of carbon dioxide (CO_2) emitted by vehicles stopped in traffic at the site. Assume that the average vehicle consumes 0.0056 gal of gas per minute sitting in traffic (based on an average of one third of a gallon per hour). Assume that each gallon burned yields 8,788 g CO_2 (8.8 kg/gal or 19.4 lb/gal).[1] Diesel is slightly higher (10.1 kg/gal), but you can use the same figure as for gasoline for your estimate, which will therefore be conservative. Determine how many vehicles idled for how many minutes in traffic for your half hour, multiply by 2, and calculate the CO_2 contributed to your block site in 1 hour that results from delays in traffic flow. Record these data into your table.

12. Measure the level of noise using a decibel meter a sound meter app on your smart phone, or estimate by subjective description (e.g., heard hum of motors, heard shouting, and jack hammering) and record these data into Table 6.8. Position yourself on the sidewalk in front of each structure for your site. Take about 1 minute to sample. In your sampling you want to capture a range if you can (e.g., when it is the quietest and when it is the noisiest).

13. If there are mixed or different land-uses at your site (e.g., houses on one end and stores on the other), note the boundaries on your sketch. Decide if other types of spatial information should be included, such as those based on socioeconomic status, as in working-class apartment building next to middle-class houses. These preliminary impressions serve to articulate a *sense of place*—part of the "understanding the problem" phase in **environmental planning**.

14. Reflect on the experience. Now that you have spent a little time examining an urban ecosystem, what do you see in common with other forms of ecosystems? Do you consider this site to have high urban metabolic function? What recommendations would you have for revitalizing the urban segment that you studied?

WRITE-UP

The write-up for this lab is a technical report. Your text and tables should be typed, along with a brief explanation of how the tables were derived. The figure you use to illustrate your streetscape

---

[1] Estimates are based on U.S. Environmental Protection Agency (US EPA). 2005. Average Carbon Dioxide Emissions Resulting from Gasoline and Diesel Fuel, EPA420-F-05-001. Available at http://www.epa.gov/oms/climate/420f05001.htm (verified 4 June 2008). Gasoline fuel weighs about 6.3 lb (2.86 kg) per gallon—less than water (8.4 lb)—but it creates much *more* than its own weight in carbon dioxide (20 lb) due to combustion, which adds the weight of oxygen from the surrounding air.

can be hand-drawn and hand-labeled or digitally created, or a combination. Your methods and results should be included along with your reflections on any benefits and limitations that you experienced in carrying out the field activity. Your report should have the following headings:

Title:

  I. **Description of Block-face**

 II. **Illustration of Block-face (Streetscape)**

III. **Urban Inputs and Outputs**
   a. **Air**
   b. **Energy**
   c. **Solid Waste**
   d. **Water and Sewer**
   e. **Traffic**

 IV. **Site Access (pedestrian and vehicle users)**

  V. **CO_2 Emission from Traffic**

 VI. **Noise**

VII. **Land-use**

VIII. **Reflections and Comparisons**

 IX. **References Cited**

REFERENCES

Adler, F.R., and C.J. Tanner. 2013. Urban Ecosystems: Ecological Principles for the Built Environment. Cambridge University Press, Cambridge, UK.

Kennedy, C., S. Pincetl, and P. Bunje. 2011. The Study of Urban Metabolism and Its Applications to Urban Planning and Design. Environmental Pollution. 159(8-9): 1965–1973.

U.S. Environmental Protection Agency (US EPA). 2005. Emission Facts: Average Carbon Dioxide Emissions Resulting from Gasoline and Diesel Fuel, EPA420-F-05-001. Available at http://www.epa.gov/oms/climate/420f05001.htm (verified 4 June 2008).

Experimental Design: Range of Tolerance

OBJECTIVES

- Demonstrate applications of the scientific method.

- Design and conduct an experiment on how abiotic environmental factors affect organisms.

- Construct a range of tolerance for a specific abiotic factor through an experiment you design and conduct.

KEY CONCEPTS AND TERMS

✓ Abiotic factors

✓ Biotic factors

✓ Confounding variables

✓ Experimental design

✓ Habitat

✓ Range of tolerance

✓ Scientific method

TIME REQUIRED: This is a two-part lab. The first part—hypothesis construction and experimental design—will take 2 hours and can be done as a homework activity. The second part of the lab—collecting data and creating a range of tolerance—will take 1.5 hours.

INTRODUCTION

Species vary in their resource needs and tolerances. For example, through evolution, some species thrive where it is hot (e.g., desert dwellers), while others thrive in the cold (e.g., Arctic species). The environment consists of **biotic** (living) and **abiotic** (nonliving or physical) parts. Within a given ecosystem, whether it is the Mojave Desert or the Arctic, each population of a species has a **range of tolerance**—the environmental conditions that a species can tolerate. (See Figure 7.1 for a sample display of a tolerance range.) Individuals within a population may also have slightly different ranges of tolerance for any environmental condition. Certain organisms are slowly changing locations due to changing environmental factors; animals are usually able to do this at a faster rate than plants, but even trees will shift their range in response to stresses. Species that have narrow ranges of tolerance are the most vulnerable to rapid environmental changes.

An organism's size, age, state of health, or genetic code can influence its tolerance range. For example, very old or young roadrunners (*Geococcyx californianus*) in the desert may be less able to tolerate extremely hot temperatures than mid-aged individuals. But there is also a limit where it is too hot for any roadrunner.

Individuals within a population of a given species respond to certain factors of their environment and seemingly ignore others. Many characteristics of a **habitat** (local environment in which an organism, population, or species lives) are variable from time to time or at different locations within the habitat. Temperature, quantity of light, moisture, soil conditions, and wind speed often vary in a forest habitat. When specific environmental factors vary continuously over a distance, a gradient exists. Light intensities can range from absolute darkness to extreme

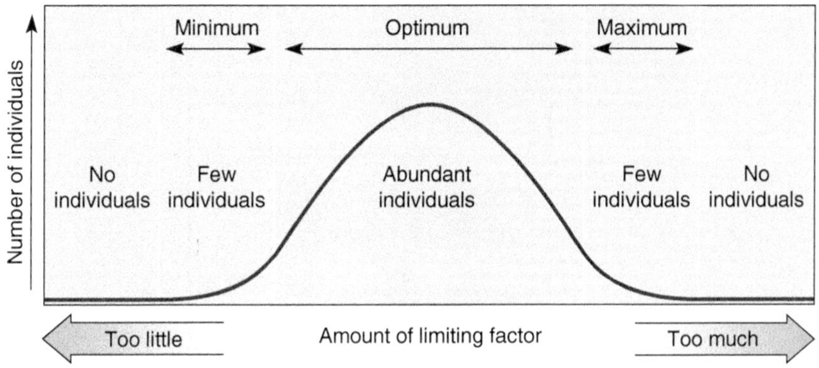

FIGURE 7.1   The range of tolerance

The limiting factor for a species in an ecosystem is any environmental resource present in excess of an organism's tolerance or in insufficient quantities to meet the organism's basic needs.

Source: Raven, P.H. and L.R. Berg. 2004. Environment. 4th ed. John Wiley & Sons, New York (p. 96).

brightness. A shady spot may be a few degrees cooler than a position in direct sunlight only a few meters away. Even at a micro level, each organism has optimal conditions to thrive. Climate change and other environmental factors will affect populations based on their tolerance. For example, as a result of heat stress and climate change, New England's sugar maples (*Acer saccharum*) produce less sugar, which means that twice as many gallons of sap needed in the 1970s are now required to make a gallon of maple syrup (Brown, 2015).

In this exercise, using the **scientific method** (see Figure 7.2), you will produce an appropriate **experimental design** (creating an experiment through which a hypothesis can be tested) to determine if a selected organism prefers certain light or heat conditions. You will then design the experiment using the equipment and materials provided, carefully collect data, and analyze the data to determine which environmental variables are significant to fruit flies or other organisms and how the organisms are influenced by key abiotic gradients. Finally, you will create a figure that depicts a simple range of tolerance for your species and abiotic factor.

MATERIALS

- Aluminum, bread pans, pie pans, or foil
- Cardboard
- Construction paper
- Fly vials (or other organism containers)
- Heating pad or plate
- Ice
- Light meter
- Light, fluorescent
- Organisms (fruit flies, daphnia, mealworms, crickets, nematodes, or other species)
- Paint brushes, fine-tipped to sort flies or other organisms
- Plastic tubes with caps, 3' long × 3" diameter
- Ring stand
- Sponge stoppers
- Tape, masking
- Thermometer

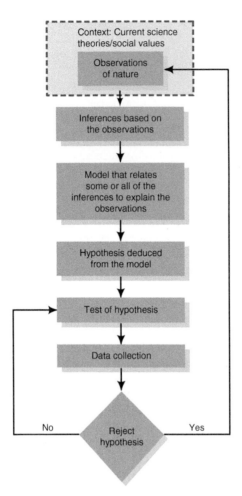

FIGURE 7.2 The scientific method

Essentially, science is a disciplined way of asking questions. Feedback in this system supports the self-correcting progress of scientific information.

Source: Botkin, D.B. and E.A. Keller. 2007. Environmental Science: Earth as a Living Planet. 6th ed. John Wiley & Sons, New York (p. 28), modified from Pease, C.M. and J.J. Bull. 1992. Bioscience. 42: 293–298.

TASKS

During this exercise, you will be divided into groups. Some groups will work with light and others will work with temperature or other variable as directed by your instructor.

1. CONSTRUCTING A HYPOTHESIS

A. Based on your assigned variable (e.g., light or temperature), go to the library or access the Internet to conduct background research on how the variable may affect your organism. You are looking for relevant peer-reviewed articles.

B. Develop an alternate hypothesis on how your organisms are affected by your assigned variable (light, temperature, etc.). Your hypothesis should focus on the optimal temperature or light.

2. DESIGNING AN EXPERIMENT

A. Design an experiment to test the hypothesis. The instructor will provide a variety of materials that can be used to construct an experimental apparatus to conduct your experiment.

- Remember, you are also trying to construct a simplified range of tolerance based on your collected data. In your experiment, you want to choose a range based on three "zones" for your assigned variable to capture an effect (e.g., too warm, optimal, and too cold a temperature).
- In designing your experiment, you need to be aware of other potential variables that can complicate your experiment. These variables are called **confounding variables**, which are variables you did not account for that can affect the outcome of your experiment. For example, if you are measuring the effects of light, excess heat from the lamp may be the variable that affects your organisms rather than the light itself, thereby undermining your results.
- You must "control" confounding variables by eliminating all potential confounding variables. Do this by setting up a control, which consists of an identical apparatus in which you replicate everything in the treatment apparatus except the variable you are manipulating. If constructed properly, the *only* difference between the experiment and the control is the treatment/variable (e.g., temperature or light).
- Remember that your experiment needs to work with the available equipment within a 1.5-hour period.
- Bring your written experimental design to the lab instructor. After it is approved, you will be provided with organisms to begin your experiment.

B. Take a photograph or prepare a sketch of your apparatus and include in your lab report.

3. COLLECTING DATA

A. Assemble and adjust your treatment apparatus so that you establish the specific gradients (e.g., the three zones) assigned to you. Be sure there also is a control. Then place your organisms in the control apparatus and the treatment apparatus. (i.e., in the control tube and the treatment tube).

B. In the control and treatment, count and record the number of organisms every 2 minutes for a total of 10 minutes and record these data into Table 7.1. Be sure to keep notes (e.g., observations, problems) throughout the experiment to include in your lab report.

C. As directed by your instructor, with new organisms, repeat the experiment (trial) two more times for a total of three trials.

4. CONSTRUCTING A RANGE OF TOLERANCE

A. Calculate the mean number of organisms for each zone and each trial.

B. Create a column graph. There should be three columns (Trial 1, Trial 2, and Trial 3) for each of the three zones on the x axis. The y axis should be the mean number of organisms per trial per zone. This graph will depict a simplified range of tolerance for your organism and your abiotic variable.

WRITING THE DISCUSSION

In writing your discussion section, be sure to address the following questions in a narrative format:

- Did you accept or reject your hypothesis and why?
- What do you think your results mean?
- What possible confounding (interfering or overlapping) variables might make one trial different from another (e.g., gender, age, stress)?
- Discuss what variables might make the control yield different experimental results from the trials. Discuss to what degree you can control these variables.
- If you were doing it again, discuss what changes you would make in setting up and conducting your experiment.

Table 7.1 Range of Tolerance Data Collection Table

Time Interval	Zone 1		Zone 2		Zone 3	
	Control	Treatment	Control	Treatment	Control	Treatment
Trial 1						
2						
4						
6						
8						
10						
Mean						
Trial 2						
2						
4						
6						
8						
10						
Mean						
Trial 3						
2						
4						
6						
8						
10						
Mean						

C. Interpret your results. Do you reject or accept your hypothesis? What were the potential confounding variables? If you were to conduct this experiment again, what would you do to improve it?

WRITE-UP

A formal experiment laboratory write-up is required for this lab using the headings below. Be sure to follow the guidelines from Part 1 of this lab manual. In writing your discussion section, be sure to address the following questions in narrative format.

Title:

I. Introduction (What was your organism? What was your variable? What background information/sources did you find to help support your hypothesis? Be sure to end this section with your hypothesis.)

II. Methods (Explain specifically what you did to test your hypothesis. Include a figure showing your apparatus.)

III. Results (Include a results table and your graphical representation of the range of tolerance.)

IV. Discussion (Did you reject or accept your hypothesis? Interpret your results. Relate your findings to the real-life environmental implications of ranges of tolerance. Identify problems with your experiment or results. Make recommendations to improve your study design.)

V. References Cited

Brown, S. 2015. Global Warming Pushes Maple Trees, Syrup to the Brink. The Plate, National Geographic. Available at http://theplate.nationalgeographic.com/2015/12/02/global-warming-pushes-maple-trees-syrup-to-the-brink (verified 07 July 2017).

REFERENCE

Experimental Design: Environmental Contamination

OBJECTIVES

- Conduct a review of scientific literature.

- Design and conduct an environmental experiment involving the effect of contaminated soil on plants.

- Use basic statistics to test a hypothesis.

- Describe the potential adverse environmental effects of contaminated soil and plants.

KEY CONCEPTS AND TERMS

- ✓ Anthropogenic
- ✓ Contamination
- ✓ Control
- ✓ Dependent variable
- ✓ Experimental design
- ✓ Hypothesis testing

- ✓ Independent variable
- ✓ LOAEL, Lowest Observable Adverse Effect Level
- ✓ Null hypothesis
- ✓ Phytotoxicity

TIME REQUIRED: This is a two-part lab. The first part—literature research, hypothesis construction, experimental design, and start of the experiment—may take 2 hours. Following a suitable growth period of 1 to 2 weeks, the second part of the lab may require 2 hours to harvest the plants, measure each plant, input data, analyze data, and then interpret the data.

INTRODUCTION **Contamination** (rendering something harmful or unsuitable by the addition of undesirable material) of soil from **anthropogenic** (human-made) pollutants is a widespread and potentially serious environmental problem. Soil can become contaminated through a variety of human activities including the use, unintentional spilling, and intentional discharge of hazardous materials and waste; agricultural pesticides and fertilizers; engineered treatment and disposal of waste; deposition of air pollutants from fossil fuel combustion; and use of salt for road deicing.

In this lab, you will be examining **phytotoxicity** (meaning poisonous to plants) by simulating, testing, and measuring the effects of contaminated soil on plants. You will be provided with seed, soil, and a contaminant. You will need to formulate a hypothesis, including the **null hypothesis**. Then, you will develop an **experimental design** and conduct your study to collect data. Finally, after you have collected your data, you will analyze the data using simple statistics to determine if you can accept or reject your hypothesis.

As depicted in Figure 8.1, you will have a **dependent variable** (the change that occurs), an **independent variable** (the variable you change/manipulate), and a **control** (everything the same except no independent variable).

Dependent variable

Independent variable

The **independent variable** is the one variable in a study you change or manipulate. You can have only independent variable per study. In this example, it is the salt.

The **dependent variable** is the change that occurs as a result of the independent variable. In this example, it is the height of the plants.

Control is the absence of an independent variable. The control must be exactly the same as the treatments because in order to measure effect, you need to compare a treatment to a control.

FIGURE 8.1   Variables in a study

MATERIALS

- Cups, 100 to 150 mL capacity or larger, 18 per study
- Fast-germinating plants such as hard red winter wheat seeds (*Triticum aestivum*)
- Masking tape
- Marker pen
- Salt, or other common household material (e.g., ammonia, antibacterial soap, dishwashing detergent, vinegar)
- Scale, in grams
- Soil, without fertilizers and additives
- Tray
- Water

TASKS

The research question for this lab is what is the lowest concentration of your contaminant that adversely affects your plant's growth? This concentration is referred to as the **Lowest Observed Adverse Effect Level (LOAEL)**. LOAELs are important in the field of risk assessment as they help set safety and/or regulatory standards for chemicals.

To determine the LOAEL, you will be conducting the following tasks:

1. Select a contaminant to test.

2. You will need to conduct background research sufficient to help you formulate a hypothesis. Your alternate hypothesis will be to predict the LOAEL (expressed as a percent of contaminant by weight in relation to the soil by weight); thus, you need some information to help you select the target concentration. For this lab, it is suggested that you start by conducting a general Internet search to find any relevant document published as *gray literature* followed by

STUDY PROCEDURES

1. Select contaminant.

2. Construct hypothesis.

3. Mark each cup based on trial and concentration.

4. In each cup, add correct weight of soil.

5. Add correct weight of contaminant for each cup's concentration, then mix.

6. Add water to fully moisten soil, then mix.

7. Add 10 seeds per cup, cover slightly with soil from within cup.

8. Place cups on a tray in proper order.

9. Place the tray in lighted area.

10. For duration, periodically check for proper moisture content.

11. Collect your data.

12. Analyze your data.

13. Interpret your data.

a search of academic literature to identify any information. Look for any previous studies regarding your contaminant and plants (remember, you are dealing with household versions of contaminants rather than industrial strength concentrations). You can also look at the manufacturer's website for relevant information. At some point, regardless of what you find, you need to make your best, informed guess; your prediction as to the LOAEL. Be sure the hypothesis is appropriate and testable and refers to the contaminant (e.g., the LOAEL of contaminant X on *T. aestivum* will be Y%). Then, formulate the null hypothesis.

3. Select five treatments (five different concentrations to test). Select your treatments so that they bracket the effect you have predicted. That is, because your goal is to find the LOAEL, select two concentrations above and two below your hypothesized concentration. For example, if your predicted LOAEL is 15%, choose 5% and 10% as the two concentrations below and then 20% and 25% as the two concentrations above.

4. Establish a control group. Remember, the control group has to be *exactly* the same as the experimental group *except* for the treatment, which is the contaminant (hence the alternative name "treatment group"). This means same cup, same amount of soil, 10 seeds, water, etc.

5. Replicate the experiment twice (for a total of three times). Thus, you should have 18 containers, and they should look like Figure 8.2.

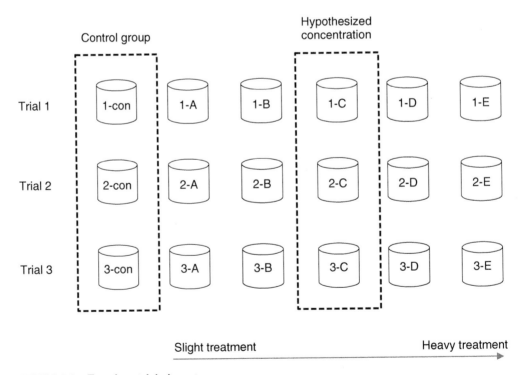

FIGURE 8.2 Experimental design setup

In this experimental design, three trials are conducted. Think of each trial as a replication. Thus, each trial must be exactly the same and each trial must have a control.

6. Depending on the plant you choose and sunlight and water, your experiment may take between 6 and 14 days. Thus, you are responsible for maintaining your experiment (i.e., do not forget to water your plants and do not overwater them). The seeds we recommend usually germinate in a few days, but extra time may be required for other seeds, dry conditions, or other factors. Should your experiment fail for any reason prior to its completion, you will have to start over—you need data to complete the lab. You may have to readjust your concentrations of the contaminant—much of science consists of learning from our so-called failures, so do not be discouraged.

7. At the appropriate time, carefully harvest your plants. Your instructor might recommend other types of data such as dry weight or number of germinating seeds. Because red winter wheat is usually a single stem, we recommend you measure the height. You will need to distinguish between the stem (green) and the root (nongreen) and measure the height of only the stem. Record the height for each plant into the data collection table (Table 8.1). Depending on the heights, record in millimeters or centimeters. *For plants that did not germinate or grow, you must input a zero.*

8. Test your hypothesis as described below. For each trial, you are comparing the mean plant growth for *each* treatment to the appropriate control (i.e., for Trial 1, 1-con to 1-A, 1-con to 1-B, 1-con to 1-C, 1-con to 1-D, and 1-con to 1-E and the same for Trial 2 and Trial 3).

Table 8.1 Data Collection Table

Trial 1						
Plant no.	Control	1-A	1-B	1-C	1-D	1-E
1						
2						
3						
4						
5						
6						
7						
8						
9						
10						
Trial 2						
Plant no.	Control	2-A	2-B	2-C	2-D	2-E
1						
2						
3						
4						
5						
6						
7						
8						
9						
10						

Trial 3						
Plant no.	Control	3-A	3-B	3-C	3-D	3-E
1						
2						
3						
4						
5						
6						
7						
8						
9						
10						

Record the height as zero if a plant did not grow

HYPOTHESIS TESTING

One of the most important elements in science is **hypothesis testing**. That is, are your results significant? If so, they will enable you to accept or reject your hypothesis with a reasonable probability of being correct. This is done through statistics—using numerical data to make comparisons and conclusions—specifically, tests for significance. This involves the application of a formal procedure for comparing observed data with a hypothesis. The results of the test of significance are expressed in terms of a probability that measures how well the data and hypothesis agree. Although we have not addressed statistics in this lab manual, you will only need to understand the basic concepts of hypothesis testing and the use of Microsoft Excel. You will be conducting a statistical test to determine whether your results are significant. That is, can you accept or reject your hypothesis?

1. *T*-TEST FOR INDEPENDENT SAMPLES

The *t*-test is the most commonly used test of significance to evaluate the differences in arithmetic means between two groups. For example, is there a significant difference between a group of patients who were given a drug and a control group who received a placebo (i.e., treatment vs. control)?

The *t*-test compares the mean of two independent variables to determine whether they are statistically different from each other. By conducting the *t*-test, you are asking whether there is a statistically significant difference between the mean height of the treatment group of a specific concentration and the mean height of the control group.

You can use Microsoft Excel to test the hypothesis. Using the data above and the two-sample "*t*-test assuming equal variance," you will get a two-tail critical value, a *p*-value, and a *t*-statistic. What does this tell you? For the purposes of this lab, we will use only the *p*-value.

The *p*-value is a measure of the probability that a difference between groups (treatment and control) during an experiment happened by chance. For example, a *p*-value of 0.05 ($p = 0.05$) means that there is a 1 in 20 chance that the result occurred by chance. (The 0.05 level is used in science as the generally minimum acceptable level of significance.) The smaller the *p*-value, the stronger the evidence against a null hypothesis. That is, it is more likely that the difference between groups was caused by the treatment. Small *p*-values suggest that the null hypothesis (i.e., no effect) is unlikely to be true. Note that the *p*-value indicates the strength of the evidence for rejecting the null hypothesis. If the *p*-value is *less than* 0.05, you can state the following: "The results are significant at the 0.05 level that X% of X contaminant by weight had a significant effect on the height of red winter wheat (*T. aestivum*)." The *t*-test tells you only whether there is a significant difference. By also looking at the mean, you can tell if a certain cup of plants is significantly taller or shorter than the control.

2. MICROSOFT EXCEL DIRECTIONS[1]

You need to compare the data (i.e., the height) in the control and the data for each treatment in each trial. Thus, you will be running four *t*-tests for each of the three trials for a total of 12 *t*-tests.

a. In Excel, enter your data from Table 8.1 for each trial separately and add a row for mean and *p*-value, which should look like this.

	A	B	C	D	E	F	G
1		Control	1-A	1-B	1-C	1-D	1-E
2							
3							
4							
5							
6							
7							
8							
9							
10							
11							
12	Mean						
13	p-value						

b. First calculate the Mean. Place your cursor in the cell below the end of the data in the control data. (In the above example, it would be cell B12). Select the f_x icon (the function command) as circled above. Look for AVERAGE, left-click OK, then make sure all the data in the column are highlighted, then select OK. Do this for the rest of your columns for each trial.

c. Now you will run the *t*-test. Place your cursor in the cell below the cell containing the Mean in the 1-A column. (In the above example, it would be cell C13.) Select the f_x icon (the function command) as circled above. Look for TTEST, left-click OK. Where it says Array 1, left-click the box with the red arrow, highlight all the data in the Control column, but do NOT include the calculated mean, hit Enter, then left-click OK. Where it says Array 2, left-click the box with the red arrow, highlight all the data in the 1-A column, but do NOT include the calculated mean, hit Enter, then left-click OK. Where it says Tails, input 2, hit Enter. Where it says Type, input 2, hit Enter. Now, left-click OK. Your *p*-value will now appear. (You will now need to do this for every cup, but you are always comparing a cup to the control, so Array 1 will always be the control data and Array 2 will vary.)

d. If the *p*-value is less than 0.05 (i.e., 5%), the result is statistically significant at the 95% confidence level. Reporting the *p*-values enables the reader to form their own conclusion about the significance of the results.

---

[1] These directions were written in 2018. Excel commands and appearances might subsequently vary but should be conceptually similar.

WRITE-UP

This is a formal experimental laboratory report:

Title:

I. **Introduction** (What is your contaminant, how much do we consume or use each year, what is it used for, what support do you have for selecting the concentration for your experiment? Be sure to end this section with your hypothesis.)

II. **Methods** (Include a figure showing your setup.)

III. **Results** (Include a results *table and a †figure.)

IV. **Discussion** (Interpretation: Relate your findings to the real-life environmental implications of your contaminant and plants. Make recommendations to refine your study in future experiments.)

V. **References Cited**

*The table presenting your results should not include the raw data. Instead, it should be similar to Table 8.2. The X% should be replaced with the actual concentrations used in your experiment.

Table 8.2 Presenting Your Results in the Lab Report

	X%	X%	X%	X%	X%
Trial 1					
Mean					
p-Value					
Trial 2					
Mean					
p-Value					
Trial 3					
Mean					
p-Value					

†The figure is different than the above table and is intended to help depict the LOAEL. See the figure below, which uses hypothetical data, as an example. Be sure to properly label your figure.

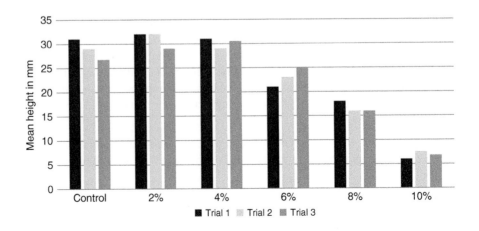

Landscaping for Energy Conservation LAB 9

OBJECTIVES

- Identify landscape design practices for energy conservation and efficiency.

- Measure and evaluate the effect of structures on wind velocity.

- Measure and evaluate the effect of shade on surface temperatures of buildings.

- Recommend changes to the external landscape to reduce energy consumption.

KEY CONCEPTS AND TERMS

✓ Climate zone

✓ Energy conservation

✓ Energy efficiency

✓ Heat loss

✓ Leeward

✓ Passive solar energy

✓ *p*-value

✓ *t*-test

✓ Windward

TIME REQUIRED: 2 hours.

INTRODUCTION

The size of the average American house has doubled since the mid-20th century. Larger houses take more energy to heat and/or cool. Yard sizes have also gotten larger; larger yards take more energy to maintain. But large yards also offer greater opportunities for using landscape design for **energy conservation** and **efficiency**. Have you noticed that parks and wooded areas during summer tend to be cooler than nearby city streets? Shading and transpiration from trees can reduce the surrounding air temperature by as much as 9°F (5°C). Trees and shrubs can be used to maximize shade during summer to cool roofs, walls, and windows. During winter, wind can increase drafts and air leakages, resulting in **heat loss** in buildings necessitating greater energy consumption. Strategically planted trees and shrubs can serve as windbreaks reducing drafts and leakage. The goal is to modify and enhance natural features that reduce the negative effects and increase the positive effects of weather (i.e., wind and sunlight) on the heating and cooling of a structure. Computer models developed by the U.S. Department of Energy indicate that the average household can save between $100 and $250 in annual energy costs through the proper placement of only three trees (see Figure 9.1).

Site considerations for energy conservation and efficiency are not limited to residential houses. Urban environments have less yard space and they act as "heat islands" with higher temperatures than in the country. In addition, because of the varying heights, shapes, and density of urban structures, winds in urban areas are quite complex, creating urban street canyons giving rise to strong localized wind. Creating sustainable buildings and communities requires the incorporation of an "ecology of design" articulated by systems thinkers such as Ian McHarg (author of *Design with Nature*, 1971). A key component of creating sustainable buildings and communities

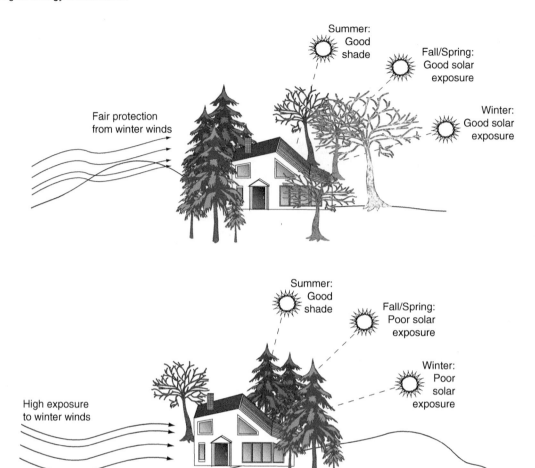

FIGURE 9.1 Landscaping for energy conservation (North is at left for both scenarios)

is to consider energy efficiency and energy conservation. Urban landscape design can reduce the heat island effect and urban canyon winds.

In this lab, you will examine a particular environmental setting—residential or commercial—and evaluate the natural and built environment regarding energy conservation and efficiency by measuring the effect of shading and wind velocity.

MATERIALS	

- Anemometer
- Calculator
- Compass
- Graph paper
- IR thermometer
- Measuring wheel or large measuring tape
- Notebook
- Wind direction Telltale
- Safety vest and helmet (if your instructor recommends them)

The following steps and questions are designed to help you prepare a technical report.

1. SITE ASSESSMENT

A. Select a site to survey for landscaping for energy conservation and efficiency. It is important for the site to have an existing structure, which can be a campus building or another building or facility approved by your instructor. What is the location of your site? Identify your site by its street address and by its climate zone. (**Climate zones** are based on temperature and moisture and used to help builders identify the appropriate building for the climate (DOE, 2017).)

B. Walk completely around the perimeter of the building. Sketch your site (building perimeter and yard/surroundings). You can do this by hand or digitally, and be sure to label features. What is the perimeter (in meters)?

C. Historically, for the general site area, from what direction do the prevailing winds blow in winter? What direction do the prevailing winds blow in summer? (The National Weather Service or the nearest airport is a source for historical data.)

2. IMPACT OF STRUCTURES ON WIND SPEED

In this step, you will be testing the theory that a windbreak impacts surface winds by testing the prevailing wind speed on the upwind side of a building (the side directly in the path of the wind) and comparing it to the downwind side of the building. If the building has an effect on the surface wind, there should be an observable difference in wind speeds. While you are testing the impact of a building, you could test any structure or natural windbreak.

A. Construct a hypothesis regarding the impact of a building or windbreak on wind speed. Below a null hypothesis is provided; construct an alternate hypothesis that is testable.

H_O: Structures have no effect on wind speed.
H_A:

B. Go to an open space near the building and determine the prevailing wind direction for that day. This can be done by observing a flag or using a telltale (a piece of string or yarn attached to a structure or stick) and using a compass. Using the anemometer, take 12 wind speed measurements 25 m from the side of the building that is in the direct path of the prevailing wind (**windward**). (Take the measurements about 1 to 2 m from each other parallel to the building.) Then, take 12 wind speed measurements 25 m from the exact opposite side of the building (**leeward**) and also about 1 to 2 m from each other parallel to the building. Record your wind speed measurements into Table 9.1 and be sure to note the units.

C. Calculate the mean, and using the *t*-**test**, calculate the *p*-**value**. State your findings (and your confidence in the findings based on the *p*-value) regarding the impact of a structure on the flow of wind.

D. For your building, based on the direction of the historical winter prevailing wind, how many deciduous shrubs and/or trees are there that could serve as a windbreak (i.e., measure the side of the structure facing the prevailing winter wind and then count the number of deciduous trees that *could* block or restrict the prevailing wind)? Mark these on your map.

E. For your building, how many coniferous shrubs and/or trees are there that could serve as a windbreak (i.e., measure the side of the structure *facing* the prevailing winter wind and then count the number of coniferous trees that *could* block or restrict the prevailing wind)? Mark these on your map.

Table 9.1   Data Collection Table, Windward versus Leeward

Sample #	Windward	Leeward
1		
2		
3		
4		
5		
6		
7		
8		
9		
10		
11		
12		

F. What conclusions can you draw regarding landscaping for wind reduction on your building using your calculations and answers to the previous questions?

G. What recommendations do you have for the site? If you recommend plantings, be sure to address their species, durability, and size and where they should be placed. Address the sustainability of your recommendations. For example, will the plants need lots of extra water or can they survive on their own? Are the plants invasive or indigenous species? Draw the recommended improvements on your site sketch. In addition to energy conservation, what are some other environmental benefits of increasing the planting of shrubs and trees?

3. IMPACT OF SHADE ON SURFACE TEMPERATURE

In this step, you will be measuring and comparing the shading impact of shrubs and trees on a structure.

A. Construct a hypothesis regarding the impact of shade on the surface temperature of a building. A null hypothesis is provided below; construct an alternate hypothesis that is testable.

H_O: Shade has no effect on the surface temperature of a building.
H_A:

B. Find a large tree that is casting a shadow on a building. Using the IR thermometer, take 12 temperature measurements on the surface of the building that is in the shade using a grid pattern (make sure the part of the building you are measuring has been in the shade for a while). Then, take 12 temperature measurements on the surface of the building that has not been in the shade using a grid pattern. Record your temperature measurements into Table 9.2 and be sure to include the units.

C. Calculate the mean, and using the t-test, calculate the p-value. State your findings (and your confidence in the findings based on your p-value) regarding the impact of shade on the surface temperature of a building.

D. Buildings designed with passive solar considerations typically have windows on the south-facing side to absorb the sun's heat energy to warm a building during winter. However, to stay cool in summer, some type of shading is needed to reduce the sun's energy from warming the building. Using your compass, find the side of the building that faces south. What is the

Table 9.2 Data Collection Table, Shade versus Nonshade

Sample #	Shade	Nonshade
1		
2		
3		
4		
5		
6		
7		
8		
9		
10		
11		
12		

estimated surface area of the windows on the south-facing side? (Measure the surface area of a window in square meters and then multiply by the total number of windows on the south-facing side.)

E. What percentage of the total south-facing side are windows (the surface area of all windows divided by the total surface area of the building x 100)?

F. What can you conclude regarding the presence of summer shade and the absence of winter shade to maximize the benefits of **passive solar energy** for your building?

The write-up for this lab is a typed technical report. Based on the subject of the report, a diagram is essential and photographs would be helpful. The following are the suggested headings for your technical report:

> **WRITE-UP**

Title: (name of the site and its location)

  I. Introduction   (why and what regarding landscaping and energy conservation)

 II. Site Assessment

III. Impact of Structures on Wind Speed

IV. Impact of Shade on Building Surface

  V. Recommendations for Improving Landscape Design for Wind and Shade

VI. References Cited

> **REFERENCES**

Department of Energy (DOE). 2017. Landscaping for Energy-Efficient Homes. Available at https://energy.gov/energysaver/landscaping-energy-efficient-homes (verified 27 June 2017).

McHarg, I.L. 1971. Design with Nature. Doubleday, Garden City, NY.

Alternative Energy: Wind Power

OBJECTIVES

- Explain how wind is used to generate electricity.

- Calculate the energy production from a wind turbine.

- Configure a basic wind farm on a small footprint.

KEY CONCEPTS AND TERMS

✓ Swept area

✓ Velocity

✓ Volts

✓ Watts

✓ Wind farm

✓ Wind park effect

✓ Wind power potential

✓ Wind turbines

✓ Yaw

TIME REQUIRED: This will take 1.5 hours.

INTRODUCTION

Electricity powers our world: heating and cooling, lights, industry, entertainment, and now our automobiles. The vast majority of electricity comes from the burning of fossil fuels. Other significant sources are nuclear, hydro, and natural gas; but, alternative energy (alternative to fossil fuels) remains a minor but increasing source of electricity. "Harnessing" wind to generate electricity is an increasingly popular proposal. Winds are generated by the sun's uneven heating of the atmosphere, irregularities of the earth's surface, and the rotation of the earth. Patterns of wind flow are modified by the earth's terrain, water bodies, vegetative cover, and structures. While wind is used to generate only 3.5% of our current electricity supply, this amount has increased dramatically in the past few years and is expected to increase as we build more *wind farms*, develop off-shore generation, and invest in household-level wind generation. (**Wind farms** are **wind turbines** grouped together into a single wind-generating "power plant" to generate bulk electrical power.)

As shown in Figure 10.1, wind turbine blades rotate due to the force produced by air moving across the surface of the blades. The rotating blades convert the energy in the wind to rotational shaft energy by spinning a shaft of an electric generator, which produces an electric current. To reduce the effects of ground terrain and structures, turbines are mounted on towers. Other equipment includes controls, electrical cables, ground support equipment, and interconnection equipment to transfer the electric current into the electrical grid similar to conventional power plants.

One of the first steps in developing a potential wind energy project is to assess the area's wind resources and estimate the available, harnessable energy. The Wind Energy Program of the U.S. Department of Energy in cooperation with the National Renewable Energy Laboratory has measured, characterized, and mapped wind resources 50–100 m above ground surface. The maps depict the annual average wind power estimates at 50 m above ground.

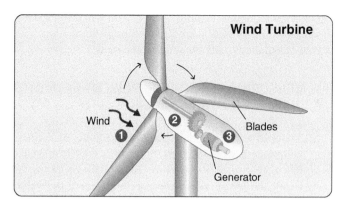

FIGURE 10.1 How a wind turbine works

To maximize the capture of steady and strong wind speeds to turn the blades, wind turbines are constructed on tall towers (e.g., 125 m) because wind currents are faster and steadier at theses heights. Thus, a tall wind turbine has greater potential to convert wind energy into electrical energy and produce more electricity at a lower cost; however, tall wind turbines are more visible and can produce aesthetic impacts and cause potential impacts to birds and bats. To transmit electricity from wind farms to the grid, new transmission lines are required, which involve cutting new swaths for the lines as well as access roads through forests, creating additional visual and physical scars. With all electricity we produce, there is a tradeoff between the economic, environmental, and social costs and benefits.

| | MATERIALS |

- Anemometer
- Calculator
- Camera
- Fan
- Masking tape
- Meter stick
- Ohms
- Paperboard, thin (manila folder)
- Protractor
- Risers (e.g., wood or plastic blocks), varying heights
- Multimeter/voltage meter
- Volts
- Wind turbine(s)

| | TASKS |

In this lab, you will perform basic measurements to determine the various factors that influence the electricity production potential of a single wind turbine. Then, using five to six wind turbines, you will construct a wind farm and determine the optimal configuration to maximize the production of electricity using quantitative measurements while minimizing its area footprint through design.

For a write-up, this lab requires a technical report. You will answer the questions and then write a summary of what you found. Make sure you use complete sentences, include diagrams or photographs, show your calculations, and always note the units.

1. MEASURING ELECTRICAL POWER PRODUCTION FROM A SINGLE WIND TURBINE

MEASURING ELECTRICITY

The three most basic units in electricity are voltage (**V**), current (**I**, uppercase "i") and resistance (**r**). Voltage is measured in **volts**, current is measured in **amps**, and resistance is measured in **ohms**.

To help you understand these terms think of a house's plumbing system. Voltage is equivalent to the water pressure, current is equivalent to the flow rate, and resistance is similar to the pipe size.

A. Set up a single wind turbine on the table. Make sure the multimeter/voltmeter is connected to the turbine (red to red and black to black). Place the turbine 30 cm away from the fan. Turn the fan on at low speed. Using the anemometer, measure and record the **velocity** in meters per second (the wind speed from the fan) directly in front of the center of the turbine facing the fan and input these data into Table 10.1.

FIGURE 10.2 Proper use of the multimeter to measure voltage

Table 10.1 Data Collection Sheet for a Single Wind Turbine

Fan speed	Distance between turbine and fan (cm)	Velocity of wind (m/s)	Voltage (V)	Power produced (W)
Low	30			
Medium	30			
High	30			

B. Next, record the **volts** and also input these data into Table 10.1. (See Figure 10.2 on using the multimeter to record volts.)

C. You measured the volts at low speed. But how much "power" is being produced? (Record the **watts** into Table 10.1). You need to convert volts into watts using the following formula:

$$P = V \times I$$

P = Power (in watts)

V = Voltage (in volts)

I = Current (in amps) (This is a given constant based on the specific wind turbine. The wind turbines used in this lab are rated at 50 mA = milliamps or 1 thousandths of an amp.) To convert to amps, we must divide this number by 1,000 = 0.05 A. Thus, the current (I) used throughout this lab for these turbines is 0.05 A.

So, if your fan is generating 3 V how much power are you generating?

$$P = 3\ V \times 0.05\ A = 0.15\ W$$

Thus, the power generated by this turbine, if the voltage meter reads 3, is 0.15 W.

   I. Based on the power produced at low speed, how many turbines would be needed to power a 60-W incandescent light bulb?

D. Turn the fan to medium speed:

   I. Measure and record the velocity (the speed of the air in front of the turbine).

  II. Measure and record the voltage.

 III. Calculate and record the watts being produced. Record your results into Table 10.1.

E. Turn the fan to the highest speed:

   I. Measure and record the velocity.

  II. Measure and record the voltage.

 III. Calculate and record the watts being produced. Record your results into Table 10.1.

F. An important question is the relationship between velocity and electrical production. Intuitively, one may believe that doubling the wind speed doubles the electricity production, but is this the case?

   I. Create a **line graph** that plots velocity and watts for your turbine (velocity is on the X axis and watts on the Y axis). There should be three data points. See Figure 10.3 for an example.

  II. Calculate the percentage difference between the watts produced at the lowest speed and the watts at the highest speed (e.g., the watts at the highest speed was 47% higher than the wind velocity at the lowest speed).

 III. In a few sentences summarize what you found regarding the relationship between wind velocity and electrical production in watts produced (i.e., did doubling the wind velocity equal a doubling of watts produced?).

CALCULATING PERCENT INCREASE

$$\frac{NV - OV}{OV} \times 100$$

where NV = New Value and OV = Old value

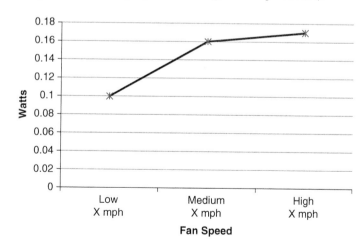

FIGURE 10.3 Example of a line graph plotting two variables, velocity and watts

G. Keep the fan set at high speed and the turbine 30 cm from the fan. Now turn the turbine 45° to the right. Record the voltage and calculate the watts. (You are now controlling the **yaw** of the turbine, which is the side-to-side movement.)

H. Keeping the fan speed and distance the same, continue to turn the turbine to the right until the blades stop spinning. Using the protractor, what is the approximate angle where the blades stop spinning?

I. Now, very slowly, turn the turbine back to the left to find the *highest* voltage output. Using the protractor, record and note the specific angle in degrees (i.e., in relation to directly facing the fan, which is 0°). Explain what happened.

J. Next, calculate the **power potential** of the **wind** at (a) low, (b) medium, and (c) high speeds from a distance of 30 cm. At each wind speed, determine the watts and record these data into Table 10.2. (One of the next steps will be to extend the blade length, but for now, measure the velocity and calculate the power potential only for normal blade length.)

Table 10.2 Data Collection Sheet for the Power Potential of a Single Wind Turbine

Fan speed	Blade length	Velocity (m/s)	Watts	Power potential (P_w)	Turbine efficiency
Low	Normal				
Low	Double				
Medium	Normal				
Medium	Double				
High	Normal				
High	Double				

You are calculating the amount of *power potential* in watts found in a specific-sized column of wind moving at a particular velocity. That is, how much potential energy is in the wind going across the turbine's blades per unit of time—the **swept area**, which refers to the *area of the circle* created by the blades as they sweep through the air. See Figure 10.4.

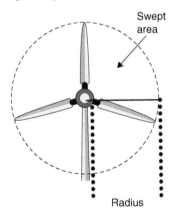

FIGURE 10.4 Windswept area and the radius

This will help you determine the efficiency of transferring that wind power potential to electricity production also measured in watts.

The formula for calculating the energy potential of wind:

$$P_w = \frac{1}{2} \times \rho \times A \times V^3$$

P_w = Wind power potential of the wind in watts

ρ = Density of the air (assume 1.0)[1]

A = Windswept area, which is the area of a circle, πR^2

π = 22/7, or 3.1416; R = radius, or the distance (in meters) from the center to the perimeter

V = Wind velocity measured in meters per second (m/s)

Step 1: Solve for A, the windswept area. For example, let's say the radius of the swept area of the turbine is 0.1 m, thus, $A = \pi R^2$ or, 3.14×0.1^2, which is 0.0314.

Step 2: Assume that you measure the wind velocity at 3 m/s. Thus, P_w would be

$$P_w = \frac{1}{2} \times \rho \times A \times V^3$$
$$= 0.5 \times 1 \times 0.0314 \times 3^3 = \mathbf{0.4239 \ W}$$

I. To calculate turbine efficiency, divide P by the wind energy potential (P_w) and multiply this by 100, which tells you how efficient (*percentage*) the turbine is at converting wind power potential into electricity. (Your answer has to be less than 100%. However, the closer to 100% the more efficient the turbine is at converting wind power to electricity.) Record your results into Table 10.2.

$$\text{Turbine efficiency} = P \ / \ P_w \times 100$$

II. Turn off the fan. Using thin paperboard and tape, add blade extenders to each blade sufficient to double their length. Recalculate P_w (note that the radius will change and thus A will change) at high speed. Now, turn on the fan, record the voltage, and calculate the new P at high speed. What can you conclude about the effect of blade extensions on wind power? Be specific (e.g., doubling the length of the blades will _____ the power production by xx%.) Record your results into Table 10.2.

2. OPTIMAL CONFIGURATION OF A SMALL WIND FARM

While individual wind turbines provide some energy, for large-scale energy production, you need a wind farm. However, it is not just the number of turbines; to produce the maximum amount of electricity, you will need to manipulate multiple variables with the goal of maximizing P_w in front of *each turbine*.

A. The first step is to set up your wind farm by carefully connecting all five to six wind turbines together as shown in Figure 10.5 (red to black and black to red). The second step is to connect the voltage meter (red to red and black to black). Turn the fan on to high speed, record the

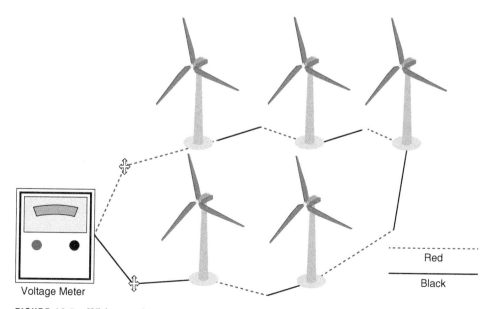

FIGURE 10.5 Wiring configuration for the tabletop wind farm

voltage, and calculate the watts produced (P)—because there are five turbines, you will need to readjust the current (I × 5).

B. Next place all five turbines directly behind each other as close as possible so that they are in a straight line perpendicular to the fan. Record the voltage and calculate the watts at the high fan speed (you will need to multiply I by 5—thus 5 × 0.05 A. Record your results into Table 10.3.

(You have witnessed what is known as **wind park effect**. Each wind turbine extracts power from the wind column. So, a turbine directly behind another has less wind power potential to turn its blades because some of the power has been extracted by the turbine in front, which is measured as P_w.)

CALCULATING PERCENT DECREASE

$$\frac{OV - NV}{OV} \times 100$$

C. To verify the wind effect, measure and record the velocity directly in front of *each turbine*. Record your results into Table 10.3. What is the percent decrease in wind velocity between the first turbine compared to the fifth turbine?

Table 10.3   Data Collection Sheet for Tabletop Wind Farm

Turbine	Velocity of wind (m/s)
1	
2	
3	
4	
5	

D. Next, spend time configuring the wind turbines in multiple ways so that you produce the highest voltage but keep them all flat on the table/bench. This will entail manipulation of placement *and* yaw. It will take a while to manipulate the turbines and their placement while taking continuous measurements. (You are looking to extract as much power as possible so you will be looking at very small incremental changes in volts.)

Once you find the optimal configuration, create a figure or take a photograph that depicts the placement of the turbines in relation to the fan and each other. Calculate the footprint of your wind farm in cm² [side (cm) × base (cm)], which is the total surface area occupied by the bases of the turbines.

E. Now imagine you have a very small footprint to construct your wind farm. You may be on very expensive land or you have access to a limited flat area on top of a hill or mountain. Or, you need to reduce the environmental impact from the construction of access roads, thus consolidating the turbines can reduce the need for roads. Your goal is to generate as much electricity as possible while placing the turbines as close as possible to each other.

In addition to manipulating placement and yaw, try varying the heights of the turbines to reduce the impact of the wind park effect. Use the provided wooden risers/blocks to vary the heights of each individual turbine—keep in mind you are trying to achieve maximum power production with the smallest possible footprint.

Create another figure or photograph that depicts the placement of the turbines in relation to the fan and each other. Determine the new footprint of your wind farm in cm². How much smaller is it than your original footprint? How much less energy was produced with the smaller footprint?

3. FUTURE ROLE OF WIND ENERGY

Based on today's lab, what are the three most important factors you learned about wind energy and its future role in serving as a reliable and efficient source of alternative energy?

The write-up for this lab is a typed technical report. Based on the subject of the report, figures and diagrams (or photographs) are essential. The following are the suggested headings for your Technical Report:

Title: (name of the site and its location)

   I. Introduction (why and what regarding landscaping and energy conservation)

  II. Single Wind Turbines

 III. Wind Farms

 IV. Role of Wind Energy

  V. References Cited

Global Climate Change and Automobiles

OBJECTIVES

- Describe the difference between samples and populations.

- Summarize the major air quality impacts of automobiles and their contribution to global climate change.

- Estimate the average annual amounts of CO_2 emitted by automobiles.

KEY TERMS AND CONCEPTS

✓ Carbon dioxide (CO_2)

✓ Greenhouse gas (GHG)

✓ Peroxyacetyl nitrate (PAN)

TIME REQUIRED: 2 hours.

MATERIALS

- Computer access
- Clipboard

INTRODUCTION

CO_2 EQUIVALENCY OF OTHER GHGs

The primary GHGs emitted from the combustion of gasoline in automobiles are carbon dioxide, methane, and nitrous oxides. However, GHGs have different potencies regarding their global warming potential. To compare this power among GHGs, they are reported as CO_2 equivalents (CO_2e). Thus, a GHG with 10 CO_2e signifies that every molecule of that gas is equal to 10 molecules of CO_2.

Carbon dioxide = 1 CO_2e

Methane = 25 CO_2e

Nitrous oxides = 298 CO_2e

Motor vehicles are a major source of anthropogenic **greenhouse gases** (GHGs), which are the primary cause of global climate change. According to the U.S. Environmental Protection Agency (2017), in 2015, the transportation sector was responsible for 27% of GHGs, which are generated through the burning of fossil fuels primarily in automobiles and trucks; ships, trains, and planes are also sources of GHGs, but comparatively a much smaller amount. As depicted in Figure 11.1, the automobile also contributes a number of other air pollutants. The health and environmental effects of global climate change are increasingly more serious. Although emissions from an *individual* car are generally low, they add up. In numerous cities across the country, the personal automobile is the single greatest polluter—emissions from millions of vehicles collectively are a major cause of poor urban air quality. Driving a private car is probably a typical citizen's most "polluting" daily activity and is the most significant contribution to environmental degradation (US EPA, 1994). The combustion of a typical gallon of gasoline releases 8.8 kg (19.4 lb) of **carbon dioxide** (CO_2).

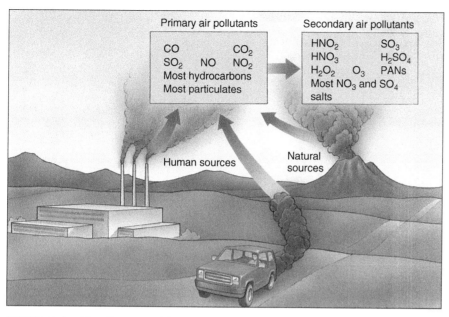

FIGURE 11.1   Air pollutants and the automobile

The automobile is a major source of air pollution. Primary pollution refers to emitted or discharged pollutants when they react in the atmosphere. Secondary pollution is produced from chemical reactions involving the primary pollutants. PAN is **peroxyacetyl nitrate**, a secondary pollutant present in photochemical smog.

Source: Raven, P.H., L.R. Berg, and D.M. Hassenzahl. 2008. Environment. 6th ed. John Wiley & Sons, New York (p. 454).

In this lab, you will be conducting basic measurements and calculations to estimate the contribution of motor vehicles to global climate change by examining fuel consumption and GHG emissions. Based on your sampling of a parking lot, you will be calculating the average fuel consumption and average GHG emissions produced by an average car at your university. You will then use this average to estimate the total contribution of GHGs and fuel consumed by your university community (students, staff, and faculty), your state, and the country.

TASKS

1. Create two alternate hypotheses regarding the fuel consumption of your car (based on miles per gallon (MPG)) and GHG emissions compared to the average car at the university. In other words, hypothesize whether the MPG of your car (you can use a family member, roommate, or lab instructor's car) is higher, lower, or the same as the MPG of the average university car. And, then hypothesize whether the GHG emissions of your car is more than, less than, or the same as the average GHG emissions based on your parking lot sample. (In your hypotheses, be sure to specify the make, model, and year of your car.)

   $H1_{MPG}$:

   $H2_{GHG}$:

2. At an assigned parking lot or portion of a parking lot, count and record the make, model, package, and year of each motor vehicle (e.g., Honda Civic, LX, 2018). To identify the year, consult Figure 11.2 on how to use the vehicle identification number (VIN) to identify the year. Record this data into Table 11.1.

   The goal is for each individual to collect data on at least 15 cars. To achieve a more robust result, you can systematically select cars, such as sample every third, fifth, or seventh car. Or you can pool your results using Goggle Sheets with the entire class to increase your sample size, but be sure to not double count the same car.

A 17-character vehicle identification number (VIN) is a unique code used by the automobile industry to identify individual automobiles that can be seen on the dashboard under the windshield on the driver's side. To tell the year of the car, look at the 10th place, which will be a letter or a digit.

| Vehicle Number | H J 3 2 N 2 C 4 5 H 9 9 1 8 7 3 1 |

In the above example, H means the year is 1987 or 2017. You should be able to make a visual determination as to 1987 v. 2017 based on the style and condition of the car.

The following are the 10th alphanumeric space codes for the model year:

Code	Year	Code	Year	Code	Year	Code	Year
A	1980	M	1991	2	2002	D	2013
B	1981	N	1992	3	2003	E	2014
C	1982	P	1993	4	2004	F	2015
D	1983	R	1994	5	2005	G	2016
E	1984	S	1995	6	2006	H	2017
F	1985	T	1996	7	2007	J	2018
G	1986	V	1997	8	2008	K	2019
H	1987	W	1998	9	2009	L	2020
J	1988	X	1999	A	2010	M	2021
K	1989	Y	2000	B	2011	N	2022
L	1990	1	2001	C	2012	P	2023

FIGURE 11.2 Using the VIN to identify the model year of an automobile

Table 11.1 Parking Lot Vehicle Tally Sheet

No.	Make	Model	Year
1			
2			
3			
4			
5			
6			
7			
8			
9			
10			
11			
12			
13			
14			
15			

3. Go to the U.S. Department of Energy's Fuel Economy website: https://www.fueleconomy.gov Select the tab "Find and Compare Cars." Under "Browse by Model," for each of your cars in Table 11.1, input the year, make, and select "Go." Select the most appropriate model from the list. If you are not sure, use your best guess. In Table 11.2, input the estimated combined city/hwy MPG. Then, for the same car, select the tab "Energy and Environment" and input the Greenhouse Gas Emissions.

Table 11.2   Gasoline Consumption and GHG Emissions for Parking Lot Vehicles

No.	Automobile (make and model)	Combined city/hwy (MPG)	GHG emissions (grams/mile)
1			
2			
3			
4			
5			
6			
7			
8			
9			
10			
11			
12			
13			
14			
15			
Mean			

4. After completing the table, input the data into an Excel spreadsheet (either as an individual or as class using a Google Sheet) to calculate the parking lot sample mean for fuel consumption and GHG emissions.

5. Using the two means, you will be calculating the total amount of fuel consumed and the total amount of GHG emissions contributed by the following. Use the average miles driven per year per driver in the United States, which is 13,476 mi (FHA, 2016).
   A. **For your university community**: Using the mean MPG and GHG emissions, what is the total amount of fuel consumed per year and how many GHG emissions are generated?[1] (The GHG emissions are expressed as grams per mile, use metric tons.)
   B. **For your state**: Using the Federal Highway Administration's website (2016) (https://www.fhwa.dot.gov), search for state motor vehicle registrations (the data will be a few years old, but will be valid). What is the total number of automobiles and noncommercial trucks registered for your state? Now, using the mean MPG and GHG emissions, what is the total amount of fuel consumed per year and how many GHG emissions are generated in your state?
   C. **The country**: Using the Federal Highway Administration's website (https://www.fhwa.dot.gov), in the section on state motor vehicle registrations, what is the total number of automobiles and noncommercial trucks registered in the United States? Now, using the mean MPG and GHG emissions, what is the total amount of fuel consumed per year and how many GHG emissions are generated in the United States?

6. Consider your car's contribution to global climate change.
   A. Describe your car (make, model, and age).
   B. What is the combined MPG for your car? Assuming 13,476 mi, how many gallons of gasoline does your car consume in a year? What are the annual GHG emissions for your car?

---

[1] You may have to consult your university department of public safety to identify how many parking permits were issued by group (faculty/staff and students).

C. Assume that everyone at the university, in your state, and in the country has the same car as your car. Using data for your car (MPG and GHG emissions), calculate the total annual amount of fuel consumed and the amount of GHG emissions generated (assume 13,476 mi per year) for the following:

  I. University or campus
   a. Gasoline consumed
   b. GHGs emitted (in metric tons)
  II. State
   a. Gasoline consumed
   b. GHGs emitted (in metric tons)
  III. Country
   a. Gasoline consumed
   b. GHGs emitted (in metric tons)

7. You have now tested your hypotheses. Can you accept or reject your hypotheses? What are some potential problems with your sample?

8. What can you conclude regarding the contribution of CO_2 from motor vehicles?

WRITE-UP

For this lab, a formal lab write-up is required:

TITLE

  I. Introduction: What did you do and why did you do it? Discuss the issues of automobiles and climate change. Also, state your hypotheses (e.g., your car compared to the sample means).

  II. Methods: How did you collect your data?

  III. Results: What did you find, be sure to include tables and figures as appropriate. (For this lab, two figures would be appropriate, one for gasoline consumed and another for GHGs emitted. On each figure, be sure to compare your car versus the average university car for the campus, the state, and the country.)

  IV. Discussion
  • Describe the potential areas of weaknesses in your conclusions (e.g., sampling, extrapolation, assumptions).
  • Briefly describe what you would do next time to improve the results.
  • Examine and reflect on this data as an individual. That is, what is the environmental significance of your findings? Were you surprised by any of the car comparisons in your data?
  • Briefly provide some appropriate policy recommendations to reduce university automobile gas consumption and the resultant air pollution.

  V. References Cited

REFERENCES

Federal Highway Administration. 2016. Average Annual Miles per Driver by Age Group. U.S. Department of Transportation. Available at https://www.fhwa.dot.gov/ohim/onh00/bar8.htm (verified 3 November 2017).

U.S. Environmental Protection Agency (US EPA). 1994. Automobile Emissions: An Overview (EPA 400-F-92-007). GPO, Washington, DC.

U.S. Environmental Protection Agency (US EPA). 2017. Inventory of U.S. Greenhouse Gas Emissions and Sinks: 1990–2015. Available at https://www.epa.gov/ghgemissions/inventory-us-greenhouse-gas-emissions-and-sinks-1990–2015 (verified 18 July 2017).

Household Contribution to Climate Change

OBJECTIVES

- Explain basic energy efficiency and energy conservation concepts.
- Compare and assess energy consumption of small home appliances.
- Describe the contribution to climate change from a typical household's energy consumption.

KEY TERMS AND CONCEPTS

✓ Carbon footprint

✓ Energy conservation

✓ Energy efficiency

✓ Hydraulic fracturing (fracking)

✓ Kilowatt hours (kWh)

✓ Lux

✓ Wattage

✓ Phantom load

TIME REQUIRED: 2 hours.

The United States remains heavily dependent on fossil fuels to generate electricity. As shown in Figure 12.1, 67.2% of electricity was produced from coal and natural gas in, 2016 (EIA, 2017a). The portion of natural gas has increased significantly especially with the use of hydraulic fracturing (fracking). (**Hydraulic fracturing (fracking)** is a process by which a high-pressure fluid containing water, sand, and chemicals is injected into a well designed to create cracks in deep-rock formations as a means to extract natural gas and petroleum.) With large supplies of comparatively inexpensive fossil fuels, the incentive to develop alternative forms of energy, or adopting policies to consume energy more efficiently through an energy conservation program, is low. Our continued reliance on fossil fuels has significant environmental, public health, and infrastructure concerns related to global climate change. In 2016, the combustion of coal in the United States accounted for 26% of all carbon dioxide (CO_2) emissions in the United States and accounted for 68% of all CO_2 emissions from the electric power sector (EIA, 2017b). Because electricity is the major source of energy consumed in homes, when carbon-intensive fuels are consumed, the household carbon footprint can be significant. (**Carbon footprint** is the total amount of greenhouse gas emissions caused by something, which can include, for example, an individual or a household.)

INTRODUCTION

POLLUTANTS EMITTED FROM COMBUSTION OF FOSSIL FUELS:

- Carbon dioxide (CO_2)
- Carbon monoxide (CO)
- Sulfur dioxide (SO_2)
- Nitrogen oxides (NOx)
- Particulate matter (PM)
- Heavy metals such as mercury

There are two choices that are not mutually exclusive: (1) find alternative energy sources (alternative to fossil fuels and nuclear) to generate electricity and (2) reduce our consumption of electricity through increased energy efficiency (**energy efficiency** is using less energy

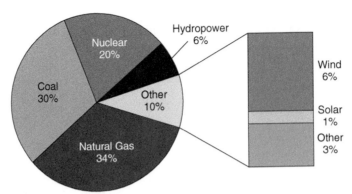

FIGURE 12.1   Electricity generation in the United States by energy source

Source: Energy Information Administration (EIA). (2017a). What is U.S. Electricity Generation by Energy Source?

to provide the same service or output) and/or energy conservation (**energy conservation** is reducing or going without a service or function that requires energy). For example, using an LED light that produces the same amount of light but consumes less electricity is an example of energy efficiency. Turning off a light when it is not being used is an example of energy conservation. Electricity generation by alternative energy is increasing significantly, especially wind. In 2016, installed wind electric generating capacity in the United States for the first time was greater than hydroelectric generating capacity, which has been the largest source of renewable electricity in the United States. As an individual, we have limited opportunities to increase consumption of alternative-energy-produced electricity in our homes, dorms, offices, and classrooms; however, as an individual, we have control over our own energy efficiency and energy conservation. One way to improve our control is to conduct an energy audit.

MATERIALS

- Calculator

- Household appliances, small (e.g., computer, refrigerator, coffeemaker, blow dryer, toaster oven, phone charger)

- Kill A Watt™ Power Meter

- Lamps (select three different types, such as fluorescent, LED, and incandescent all with same power rating/watt equivalence)

- Light meter

TASKS

In this lab, we will perform some basic components of a home energy/carbon emissions audit by focusing on electricity consumption. You will be measuring the electricity consumption of different forms of lighting and of basic household appliances. You will assess and compare the efficiency of different lights. You will then calculate the annual amount of CO_2 generated per appliance. The final result will be for you to estimate the energy consumption and associated CO_2 generation of a typical household.

Remember to note the correct units for each number.

1. ELECTRICAL LIGHTING AND CO$_2$

Lighting is a significant source of energy consumption. In the typical American home, lighting accounts for about 5 to 10% of total energy use. If extensive outdoor lighting is used, the annual lighting consumption will be significantly higher. The consumption of lighting can be reduced if more efficient lighting is used (and, of course, energy conservation measures are adopted). For this section, you will compare the efficiency of three types of lights.

A. In a test box or darkened room, place a light meter on the bottom. Next, connect an LED lamp to the Kill A Watt meter, which is then connected directly to an electrical outlet. This meter measures the actual electricity consumed by the lamp. Turn on the LED light, and with the light meter, record the amount of **lux** produced. Using the Kill A Watt meter, record the amount of energy (watts) consumed. Input these results in Table 12.1. To determine the efficiency of the LED light, divide the amount of lux by the **wattage**, which provides a lux per watt output. Input your data into Table 12.1.

B. Replace the LED lamp with the fluorescent lamp. Measure the amount of light coming from the fluorescent lamp, record the lux, then measure and record the energy consumed. Calculate the lux per watt produced. Input your data into Table 12.1.

C. Replace the fluorescent lamp with the incandescent lamp. Measure the amount of light coming from the incandescent lamp, record the lux, then measure and record the energy consumed. Calculate the lux per watt produced. Input your data into Table 12.1.

D. All three lamps should have the same rated wattage but which lamp is *the most efficient* in providing light?

E. Based on your measurements, how much more efficient the most efficient lamp is than the least efficient one? Give your answer in terms of percent difference, and be sure to identify the lamp type.

F. Multiply the wattage of each lamp by the number of hours it is used each year (assume 4 hours per day and 300 days per year) and then divide by 1,000 to get the **kilowatt-hour (kWh)** usage per year (**kWh** is the standard charge to residential and commercial customers). Multiply the kWh by the current price of residential electricity in your state (or use the current U.S. average residential retail price for electricity, which in 2016 was 12.7 cents, or $0.127 per kWh). Input this data into Table 12.1. How much money would you save if you (1) replaced the least efficient lamp with the most efficient lamp?

> **CALCULATING THE PERCENT DIFFERENCE**
>
> $$\frac{\text{New Number} - \text{Original Number}}{\text{New Number}} \times 100$$
>
> For example:
>
> Lamp A = 30 lux/watt
>
> Lamp B = 12/lux/watt
>
> $(30 - 12)/12 \times 100 = 150\%$

G. You calculated the difference for only a single lamp. What about the savings for an entire house? On average, a three-bedroom house has 60 light sockets. (1) Assuming all lamps in the house are incandescent lamps (60), what would be the total annual savings if all 60 lamps were replaced with the energy-efficient LED bulbs (assume that the replacement would be with the same wattage equivalent)? (2) What would be the total savings in a small city with 10,000 three-bedroom houses where all incandescent lamps were replaced with LED lamps?

Table 12.1   Comparison of the Energy Efficiency of Standard Lamps

	LED	Fluorescent	Incandescent
Watts			
Lux			
Efficiency (lux per watt)			
Annual cost			

2. HOUSEHOLD APPLIANCES AND CO_2

For this section, assume your home gets electricity generated from a coal-burning power plant. When operating, each electric household appliance would indirectly be responsible for emitting a certain amount of CO_2. By measuring how much CO_2 each electrical appliance is emitting, you can begin to see how energy efficiency and energy conservation, if implemented on a wide scale, could have an effect on global CO_2 emissions and thus global climate change. For this portion of the lab, you will be calculating the number of kilowatt-hours (kWh) an individual electrical appliance uses and then calculating how many pounds of CO_2 would be emitted. (You also can calculate the annual cost of each appliance to calculate the economic benefit of energy efficiency and energy conservation.)

A. Select a number of electrical home appliances commonly found in a home or dorm such as computer, refrigerator, television, coffeemaker, blow dryer, toaster oven, phone charger. Plug each appliance directly into a Kill A Watt™ Power Meter. Turn on the appliance and operate it at full or high speed if available. For each appliance, record the watts in Table 12.2.[1]

B. For each appliance, determine if it is drawing a phantom load (also known as standby power or vampire power) and how much power is consumed as phantom load. **Phantom load** is the electric power consumed by electronic and electrical appliances while they are turned off (roughly equivalent to a leaky faucet).

C. Estimate the number of hours each appliance might be used in an average household or dorm room during a typical day and record this in Table 12.2. Then, calculate and record the annual kWh consumed by each appliance in Table 12.2 using the following formula:

Annual kWh = watts consumed × hours per day × days per year/1,000

So, for an appliance drawing 100 W used for 8 hours per day and 300 days per year:

$$100W \times 8h \times 300D/1,000 = 240 \text{ kWh}_{YEAR}$$

D. Next (1) calculate the amount of CO_2 *by weight* produced for each appliance and record in Table 12.2 and (2) then sum the total to estimate the total CO_2 produced from electrical appliances for one household. (2.8 lb of CO_2 are produced for every kWh consumed if electricity was generated by coal.)

Table 12.2 Single Family Home Contribution to Carbon Emissions

Appliance	Watts	Hours/day for this appliance	Days/year for appliance	Annual kWh consumed	Annual CO_2 emitted (lb)
TOTAL					

[1] Some Kill A Watt™ meters may not record watts but record kilowatts (kWh), which is a cumulative measure. Thus, start your appliance and record the time. At 5 minutes, note the kWh. Multiply this by 12 to obtain kWh per hour of operation. You need to unplug or reset the meter every time to set the time and kWh back to zero.

E. *Optional*: For each appliance, calculate the individual cost and total annual cost based on how many hours and days per week each is used. For each appliance, identify one energy conservation action and one energy efficiency action to reduce the electricity consumption of each appliance while still enjoying the services provided by the appliance.

3. HOUSEHOLD CONTRIBUTION TO CLIMATE CHANGE

A. In the previous section we assumed that the electricity supplied to the dorm or home was generated from a coal-burning power plant. In some states, however, very little coal may be used to produce electricity and instead a combination of sources is used including renewable energy (e.g., wind power, hydropower, and solar), natural gas, nuclear, and trash incineration. Go to https://www.eia.gov/electricity/state and select your state. What is the primary energy source for electricity generated in your state? What rank is it? What rank is your state regarding emissions of CO_2? How would this information change the calculations of your CO_2 emissions?

B. If your state shifted toward more alternative energy sources (e.g., wind or solar that does not directly emit CO_2) for electricity by, for example, 25%, how would this change your answer in 2A regarding CO_2 emissions related to electrical appliances used in your home or dorm?

The write-up for this lab is a technical report. Use the following headings format for your report. The tone of your report should be neutral and fact-based.

| WRITE-UP |

Title:

  I. Introduction (what you did, when you did it, why you did it, and where you did it)

  II. Background Research (discuss the importance of the lab's topic in relation to environmental science)

  III. Electrical Lighting and CO_2

  IV. Household Appliances and CO_2

  V. Looking at the Whole Energy Picture (address energy efficiency and energy conservation)

  VI. References Cited

| REFERENCES |

Energy Information Administration (EIA). (2017a). What is U.S. Electricity Generation by Energy Source? Available at https://www.eia.gov/tools/faqs/faq.php?id=427&t=3 (verified 18 July 2017).

Energy Information Administration (EIA). (2017b). Where Greenhouse Gases Come From. Available at https://www.eia.gov/energyexplained/index.cfm?page=environment_where_ghg_come_from (verified 18 July 2017).

Hydrology and Groundwater Pollution

OBJECTIVES

- Using a groundwater model, describe and demonstrate how an artesian well works, how an injection well works, and how groundwater can become contaminated due to an underground storage tank.

- Experiment and make inferences from a physical groundwater model about groundwater flow in aquifers and various soil horizons.

KEY TERMS AND CONCEPTS

✓ Aquifer

✓ Artesian aquifer

✓ Artesian well

✓ Contaminant plume

✓ Darcy's Law

✓ Groundwater

✓ Hydraulic pressure

✓ Hydrology

✓ Permeability

✓ Porosity

✓ Plume

✓ Water table

✓ Zone of saturation

TIME REQUIRED: 1 hour using the groundwater model (if using an existing model) and 30 minutes for cleanup to flush coloring out of the groundwater model.
Add 1 hour if building your own groundwater model in a beaker, bottle, or jar.

INTRODUCTION

There are many uses of groundwater, but by far the biggest use is for irrigation, which ultimately affects all of us through our food supply (Groundwater Association, 2017). **Groundwater** is also a major source of drinking water as about half of the population in the United States obtains its drinking water from groundwater. Water cycles through nature, entering the ground through recharge areas. Water is stored in underground reservoirs called aquifers. An **aquifer** is almost always a layer of rock or soil, with water held in the spaces between the soil or rock particles or fractured rock (Figure 13.1). Some states are rich in groundwater, but in many states, aquifer depletion is a serious problem (Figure 13.2).

Groundwater can be contaminated by agricultural pesticides and fertilizers, septic systems, above-ground storage tanks, underground storage tanks (USTs), landfills, contaminated surface waters, wells, chemical spills, and through saltwater intrusion. Cleanup of contaminated groundwater depends on the understanding of how groundwater flows. Figure 13.3 shows how a leaky underground fuel storage tank can be cleaned up by pumping the contaminated material out of extraction wells. The well placement sites and depths are based on a computer model of how the groundwater and contamination plume is predicted to behave. In making the prediction, we need to know the underground conditions, including a description of the various subsurface layers and their characteristics. A layer with high **permeability** allows water to

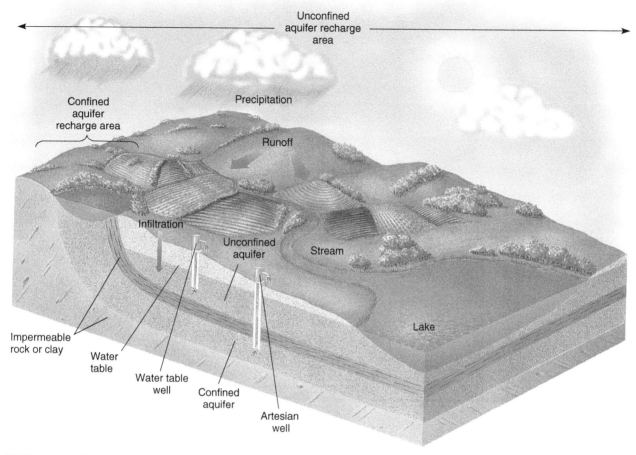

FIGURE 13.1 Groundwater

Excess surface water seeps downward through soil and porous rock layers until it reaches impermeable rock or clay. An unconfined aquifer has groundwater recharged by surface water directly above it. In a confined aquifer, groundwater is stored between two impermeable layers and is often under pressure. **Artesian wells**, which produce water from confined aquifers, often do not require pumping because of pressure.

Source: Raven, P.H., L.R. Berg, and D.M. Hassenzahl. 2010. Environment. 7th ed. John Wiley & Sons, New York (p. 305).

move quickly through and between it due to its high **porosity**: large amount of open spaces (pores) between soil particles. Sand has high permeability and clay has low permeability. The location of the water table is also a factor. The **water table** is the upper portion or boundary of the saturated areas of the earth (**zone of saturation**). If an area is not saturated with water, as in the case of being above the water table, then the contamination may sink until it reaches the water table.

In this laboratory activity, you will use a groundwater flow model (Figures 13.4 and 13.5) to experiment with and observe the behavior of water and contaminants (dyed water for visibility) in the water (**hydrology**). **Hydraulic pressure** (the forces that cause water to move) can reveal the way the water and contaminants respond in simulated underground rock formations, streams, lakes, leaky wells, and storage tanks. The model allows experimentation with aquifers, permeability, porosity, water tables, and water flow (hydraulic) conditions that you can manipulate.

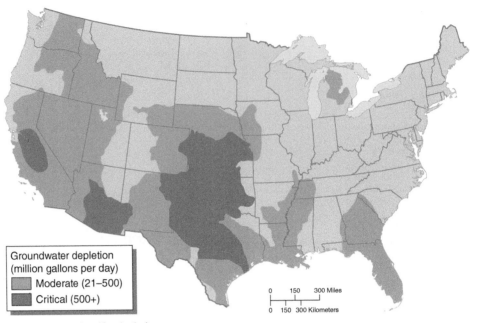

FIGURE 13.2 Aquifer depletion

Aquifer depletion is a widespread problem in the United States, particularly in the High Plains, California, and southern Arizona (Adapted from Englebert, E.A. and A. Foley).

Source: Raven, P.H., L.R. Berg, and D.M. Hassenzahl. 2010. Environment. 7th ed. John Wiley & Sons, New York (p. 313).

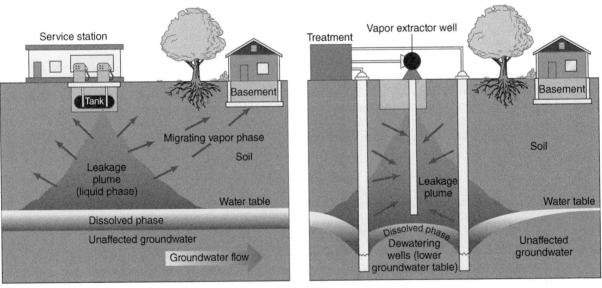

FIGURE 13.3 Underground water contamination

Diagram illustrating (a) a leak from a buried gasoline tank and (b) possible remediation using a vapor extraction system. Notice that the liquid gasoline and the vapor from the gasoline are above the water table; a small amount dissolves into the water. All three phases of the pollutant (liquid, vapor, and dissolved) float on the denser groundwater. The extraction well takes advantage of this situation. The function of dewatering wells is to pull the pollutants in where the extraction is most effective.

Source: Courtesy of the University of California Santa Barbara Valdose Zone Laboratory and David Springer. Botkin, D.B. and E.A. Keller. 2011. Environmental Science: Earth as a Living Planet, 8th ed. John Wiley & Sons, New York (p. 412).

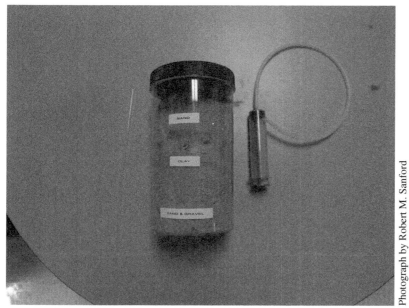

FIGURE 13.4   Improvised groundwater model

You can make your own low-cost model out of a jar or other clear container, gravel, clay, sand, a straw, and a plastic syringe.

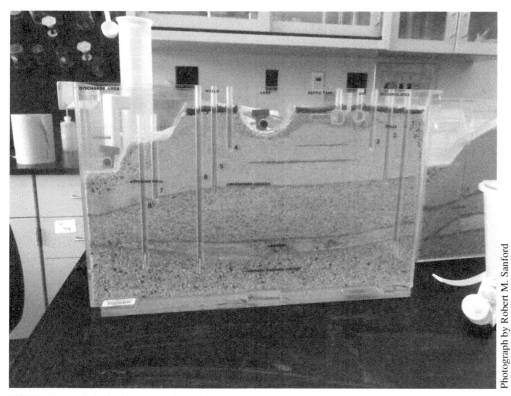

FIGURE 13.5   A typical groundwater model

These models can be purchased or made locally. The configuration need not be exact and can be modified to replicate local geological conditions. In this case there is a septic tank, UST, lake, stream, artesian wells, injection wells, recharge and discharge areas, and a stratigraphic profile of various layers. A reservoir in the back of the model holds water that can be drained from the lower portions of the model, adding pressure to move water through the system.

MATERIALS	• Plastic syringes

• Water, enough to fill your model, and to rinse it later

• Food coloring or colored drink powder such as KOOL-AID®

• Groundwater model

You can make your own model:

• 2-L clear soda plastic bottle, or clear similar size glass or clear plastic jar, beaker, or other container.

• Clay (you can use clay-based kitty litter or modeling clay from a hobby store, or clayey soil). (Choose enough to make a layer of about 1 cm thick in your container.)

• Pea-gravel or aquarium stones to make one or more layers totaling about 8 to 10 cm thick.

• Peat moss or cut up a green pot scrubber. This will be the thin top layer of your model.

• Ruler to measure the layers of your containers.

• Sand to make one or more layers totaling approximately 10 cm thick.

• Small plastic syringes—you can get these from a pharmacy or hardware store.

• Three or four transparent drinking straws or use three to four transparent plastic tubes of similar thickness.

• Two different food coloring, two different colors, or colored drink packages (e.g., KOOL-AID®).

• Water (enough to fill your container).

If you are making your own groundwater model, you can find some guidance and advice online (e.g., Millen, 2011; Brunei, Todd, and Johnson, 2013).

TASKS	1. The first step is to add enough water to your groundwater model to bring it within about 1 cm of the top layer (unless this has already been done). Now you are ready to begin experimenting with your model in accordance with the items below.

2. An **injection well** is intended to inject water (along with anything in that water) into the ground beneath underground sources of drinking water. Injection wells are primarily used to dispose of liquid material. Sometimes, a regular water source (noninjection) well can also be contaminated. Select an injection well site (or if using a homemade model, use a straw as your injection well) and use a syringe filled with colored water to represent contamination. Insert enough colored water to completely fill the injection well. Now, observe what happens to this "plume" of contamination. (A **plume** is a volume of contaminated groundwater that extends downward and outward from a specific source.) How does it move or spread? How long did it take to move or spread? How does the plume's size and movement depend on the characteristics of the layer in which it is released?

3. An **artesian aquifer** is groundwater that is under positive pressure because it is a confined aquifer, which means it pushes water up and away, so any well tapped into it would have water coming up under pressure, making it an artesian well. Using an empty syringe, press the plunger down, then insert it into the well and pull out the plunger so the artesian well site draws up (extracts) water from the model. Discard the water and continue drawing up by pulling out the syringe plunger to apply suction while watching your contamination source from the injection well. Does the contaminated source create a plume that is pulled from the

injection well toward your extraction well? (Note: if you do not have an artesian well, then use any other well.) Describe your observations of how the contamination plume behaves.

4. If you have a groundwater model with a representation of a stream or lake in it, does the water color change in the river or stream as a result of performing items 1 and 2? Describe your results.

5. If you have a model with an UST or septic system tank, use your syringe to fill the tank with a different colored water and observe what happens when the UST leaks. Then go to your artesian well and draw up from it so that you are "loading" the system with vacuum pressure. Does this pressure increase the movement of the **contaminant plume** out of your UST or septic system? Describe the behavior. Does the contaminant mix with that of the injection well's plume?

6. Examine your system to determine where the contaminant plumes flow in relation to your layers of sand, clay, gravel, and other materials as a result of your experiments above. To what degree does the clay act as a barrier? Does water (and associated contamination) move faster through the more porous gravel?

7. Simulate a rain event by adding water to the top of your system. You will have to experiment to see how much water represents a heavy rain event or your instructor may assign a specific amount. What did you observe with the seepage of water and movement of the contaminant plume?

8. Design your own experiment. For example, you might simulate drought conditions and see how this affects the contaminant plume. Or simulate a 100-year flood event by dumping a large amount of water onto the top of your model, or add more contamination. Carry out your experiment and report the results.

9. How does Figure 13.1 compare this with the groundwater where you live? You may have to look up information about groundwater in your area to answer this. (Each state has a State Geologist and water resources department of some sort that should be a good reference for this information.) What soil and rock layers and other subsurface environmental conditions characterize your area? For example, if you live in a desert environment, your groundwater may be very deep with a low recharge rate.

10. Think about your groundwater model experimentation. How has it helped you to think about specific pollution prevention measures that could protect groundwater?

1. BONUS OPTION: DEMONSTRATION OF DARCY'S LAW

Darcy's Law describes the relationship of fluid in porous medium (in our case, water in an aquifer). The flow rate, Q = the permeability coefficient K × the cross-sectional area (A) × the difference between the heights of the flow inlet head (h_1) and outlet head (h_2), all divided by the path length (L) of the flow. Thus, $Q = KA(h_1 - h_2)/L$. If you use different outlet points, you can determine the pressure by starting with a known flow rate for water to enter the model. Demonstrate Darcy's Law and report the pressure. Label and show all work in doing your calculation.

For this lab, a lab write-up in the form of a technical report is required. Use the following headings:

| WRITE-UP |

Title:

I. **Introduction:** Describe how your experiment was set up, including what model you used and how it was configured.

II. **Findings:** Answer items 1 through 9 in a narrative fashion with complete sentences.

III. **Discussion:** Answer number 10 above.

IV. **References Cited**

REFERENCES

Brunei, D., B. Todd, and P. Johnson. 2013. Modeling Groundwater Flow Experiment. American Society for Engineering Education. 2013 ASEE Southeast Section Conference [Online]. Available at http://se.asee.org/proceedings/ASEE2013/Papers2013/145.PDF (verified 22 June 2017).

Groundwater Association. 2017. What Is Groundwater? Available at http://www.groundwater.org/get-informed/basics/groundwater.html. (verified 24 July 2017).

Millen, K. 2011. Make Your Own Ground Water Model. National Ground Water Association (NGWA) [Online]. Available at http://www.ngwa.org/Fundamentals/teachers/Pages/Make-Your-Own-Ground-Water-Model.aspx (verified 20 June 2017).

Stormwater Generation and Management

OBJECTIVES

- Describe how stormwater runoff is generated and how it impacts the environment.

- Apply the scientific method to measure and analyze the variables that cause stormwater runoff and contaminated stormwater runoff.

- Identify best management practices (BMPs) to reduce the environmental impacts of stormwater runoff.

KEY CONCEPTS AND TERMS

✓ Best management practices (BMPs)

✓ Combined sewer overflow (CSO)

✓ Hypothesis

✓ Impervious

✓ Nonpoint source pollution

✓ Permeability

✓ Stormwater runoff

INTRODUCTION

As human development improves or expands with the construction of buildings, roads, sidewalks, and parking lots, we also create impermeable barriers that cover the land. As a result, when precipitation events occur, the generation of stormwater runoff increases (**stormwater runoff** is the water from rain or melting snow that "runs off" the land instead of seeping into the ground). This runoff can carry pollutants and can negatively impact receiving waters (e.g., rivers, streams, wetlands, lakes, and ocean). Or, there can be too much runoff for a receiving water, potentially causing flooding. Best management practices can be employed to reduce the generation of stormwater and/or reduce the presence of pollutants in stormwater runoff.

Stormwater runoff is a natural occurrence. As shown in Figure 14.1, the occurrence and volume of runoff is a function of the amount and ability of precipitation to infiltrate the ground. Runoff results from too much precipitation over a short time period, saturated ground, or when **impervious** surfaces prevent infiltration. Development within watersheds can cover the natural surface with impervious surfaces such as roofs, driveways, parking lots, sidewalks, and roads. The impervious surfaces block infiltration of precipitation, leading to more stormwater runoff. Runoff usually flows into the nearest storm drain or water body (e.g., wetland, brook, stream, creek, canal, river, pond, lake, or ocean). An additional issue associated with impervious surfaces is that stormwater can "collect" pollutants on these surfaces (see Figure 14.2). Thus, commonly applied substances such as pesticides and fertilizers can become pollutants and enter a nearby water body. Further, Other pollutants, such as oil, soaps, salt, and pet waste, which have accumulated during dry periods, can then become pollutants in stormwater. Stormwater running off from exposed surfaces (e.g., construction sites, agricultural lands, and bare soil) can cause erosion and then sedimentation in water bodies, adversely affecting aquatic organisms and habitats. If the

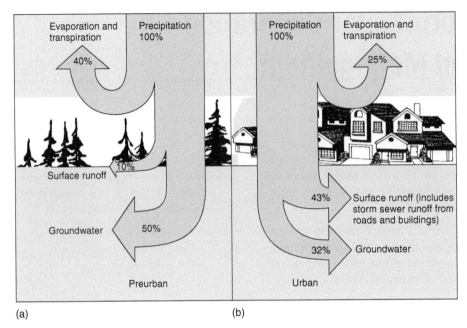

FIGURE 14.1 Example of infiltration and stormwater runoff of precipitation

Urbanization can have a significant effect on infiltration and stormwater runoff as urbanization increases the number of impervious surfaces.

Source: Raven, P.H., L.R. Berg, and D.M. Hassenzahl. 2008. Environment. 6th ed. John Wiley & Sons, New York (P. 314).

FIGURE 14.2 Example of contamination of stormwater runoff

Source: Raven, P.H., L.R. Berg, and D.M. Hassenzahl. 2008. Environment. 6th ed. John Wiley & Sons, NY (P. 518).

runoff is not channeled, it is considered to be **nonpoint source pollution**. Runoff from point sources and nonpoint sources needs to be managed.

Stormwater is an issue for municipalities that rely on a combined sewer overflow system. (A **combined sewer overflow (CSO)** is a sewage collection system that is designed to collect sewage, stormwater, and industrial wastewater in the same system for subsequent treatment.) This system works well during low precipitation events as contaminated stormwater can be collected and treated before discharge. However, if too much water, wastewater, and sewage enters the system at the same time, significant amounts of untreated sewage and untreated stormwater can be discharged directly into a receiving water. A root cause is that the sewage treatment plant may have been built many years ago; increased population and increased development can exceed the original design capacity of the plant.

MATERIALS

- Bucket, filled with 6 to 8 L of water

- Compass (or Smartphone compass app)

- Container, 1-L

- Internet access

- Map, topographic or Google Earth

- Measuring tapes, 15 to 25 m

- Meter sticks or m² sampling quadrant—Notebook

- Pen

- String

- Stakes and chalk to mark/plot where to dump water at the study area

- String levels

TASKS

In this lab, you will be conducting a series of experiments to assess the potential for stormwater to be generated, its potential for picking up pollutants, and its likely flow.

1. GENERATING HYPOTHESES

Consider the following questions:

- If stormwater cannot be absorbed into the ground because of an impervious surface, can it create a greater volume of runoff?

- Can stormwater runoff "collect" pollutants from surfaces?

- If there is a downward slope, will this impact the volume or flow of stormwater runoff?

- Will the direction of the downward slope guide the flow of stormwater runoff?

The first step is to generate suitable hypotheses to answer the above four questions. A **hypothesis** is a statement of the cause (independent variable) *and* effect (dependent variable) in a specific situation. It is an educated guess presented as a statement that predicts the outcome of an experiment that is based on a review of the relevant literature or observations. Your hypothesis must be testable, must be unambiguous, and must have a dichotomous answer (it has to be answered by a yes or a no).

In designated teams, develop and write *two* tentative alternate hypotheses (two for *each* of the following) regarding the impact of stormwater runoff on your campus. **Permeability**, which is the ability of a substance to allow water (or other liquid) to pass through it.

A. The permeability (*cause*) of a surface and stormwater generation (*effect*).
   i. H_A1:
   ii. H_B2:

B. The type of uses (e.g., parking) of a particular surface (*cause*) and the contamination of stormwater (*effect*).
   i. H_A1:
   ii. H_B2:

C. The relationship between slope (*cause*) and the volume of stormwater runoff (*effect*).
   i. H_A1:
   ii. H_B2:

D. The relationship between the compass direction of the downward slope (*cause*) and the direction (N, S, E, or W) of the flow of stormwater runoff (*effect*).
   i. H_A1:
   ii. H_B2:

Circle the best hypothesis for each question for a total of four hypotheses and then present them to your instructor for approval.

2. DATA COLLECTION

For this lab, you will have two study areas, each comprising three sample plots. The two study areas are the Parking Lot Study Area that will serve as the impervious surface area *and* Green Space Study Area (a nonparking lot, such as a lawn, field, or other vegetative area) that will serve as the pervious surface area.

A. Describe the type and likely permeability of the surface of each study area, note its relative location and surrounding, and provide an estimate of the surface area in square meters using pacing, measuring tape, or other measuring device.

> NOTE: For each sample plot, you will be measuring the amount of pollution and then will be calculating the amount of stormwater runoff. Conduct each task simultaneously for each sample plot.

B. At *each* study area, randomly select three sample plots (in this case, a plot is a specific location where you will collect your data). For each sample plot, construct a (1 m^2) plot by using a meter stick to construct a square or using a m^2 sample quadrant. For the Parking Lot Study Area, mark out the plot using chalk, and use stakes for the Green Space Study Area. For each sample plot, list the type *and* amount of each category of pollutants you see (e.g., percent coverage by sand, salt, oil, fluids from automobiles, litter, and cigarette butts) and record into Table 14.1.

C. At each sample plot, from a height of 1 m, dump 1 L of water (dump the 1 L of water over a 5-second period) into the center of the plot. At 10 seconds, estimate (%) how much of the water runs off the surface (i.e., did not infiltrate into the ground) and record this in Table 14.2. Repeat this for a total of six sample plots, three at each study area.

D. You will now estimate how much water would run off each study area during a 1-in rainstorm. Use your answer from C and the below information to answer this question.
   I. How large is the Parking Lot Study Area? ___ m^2 The Green Space Study Area? ___ m^2
   II. There are 0.017316 qt in 1 in^3; there are 1,550 in^2 per 1 m^2 (and of course there are 4 qt in 1 gal.)

Table 14.1 Pollutants Found at Each Study Area

	Parking Lot Study Area			Green Space Study Area		
	Plot A	**Plot B**	**Plot C**	**Plot A**	**Plot B**	**Plot C**
Types of pollutants						
Amount of pollutants						

Table 14.2 Percentage of Runoff at Each Study Area

	Parking Lot Study Area				Green Space Study Area			
	Plot A	**Plot B**	**Plot C**	*Mean*	**Plot A**	**Plot B**	**Plot C**	*Mean*
Percent of water as runoff								

    III. So, 0.017316 qt × 1,550 in² /4 qt = ___ gal of water per m². Multiply this volume (gals per m²) and the total area (m²) of each site to get total gallons for each study area.

        a. __ gal per m² × ___total m² area = ___ gal of precipitation falling on the Parking Lot during a 1-in rainstorm.

        b. __ gal per m² × ___total m² area = ___ gal of precipitation falling on the Green Space during a 1-in rainstorm.

    IV. The mean percentage of runoff for the Parking Lot site (from Table 14.2) is __. The mean Green Space site percentage is ___.

    V. Multiply IV and III for each site to determine the volume of potential stormwater runoff in gallons for the Parking Lot (__ gal) and the Green Space (___ gal).

    VI. What factors did you observe that could impact your answer for V (e.g., season, humidity)?

E. As shown in Figure 14.3, you will be calculating the mean slope (the mean percent slope and the mean direction of the slope) for both study areas by taking *three measurements* at each.

    I. To calculate the mean percent (%) slope, use the string, string level (or Smartphone), meter stick, and measuring tape and take three measurements over a 15 to 25 m distance by following Figure 14.4. The three measurements should be from the center of each sample plot. Record the results in Table 14.3.

    II. To calculate the mean direction (i.e., compass direction) of the slope, using a compass (or Smartphone) determine in specific degrees (e.g., 260°) the *direction of the downward slope*. (In other words, on average, which direction would water tend to flow based on your three measurements?) The three measurements should be from the center of each sample plot. Record the results in Table 14.3.

3. DETERMINING THE POTENTIAL RECEIVING WATER

A. Back in the lab or library, examine the Google Earth map of the campus. Based on your measurements, mean slope, and mean compass direction of the slope:

    I. What is the name of the nearest surface water body most likely to be the receiving water of any stormwater generated from your specific study area?

    II. How far is it in meters from the study area?

    III. What compass direction (in degrees and cardinal direction, NW, SE, NNE, etc.) is it from the study area?

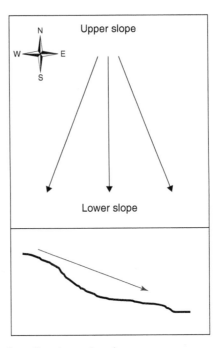

FIGURE 14.3 Measuring the slope direction and angle

From the upper slope (the high point), take three measurements to determine the mean slope and the mean compass direction of the slope.

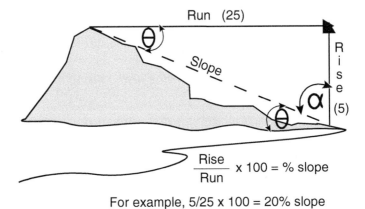

$$\frac{\text{Rise}}{\text{Run}} \times 100 = \% \text{ slope}$$

For example, 5/25 x 100 = 20% slope

% slope	∠ slope
20%	11.3°

$$\alpha + \theta = 90°$$

$$\text{Tan } \theta = \frac{\text{opposite}}{\text{adjacent}} = \frac{5}{25}$$

$$\text{Tan } \theta = .2$$

$$\theta = 11.3°$$

FIGURE 14.4 Illustration of how to determine the slope at each sample area

Table 14.3 Slope and Direction at Each Study Site

	Parking Lot Study Area				Green Space Study Area			
	#1	#2	#3	Mean	#1	#2	#3	Mean
Slope (%)								
Direction of slope (°)								

B. Now that you have completed the observations and data collection, you need to analyze your data and conduct the comparisons between the two types of surfaces and state clearly a response to each of your four hypotheses:

    I. Hypothesis 1: do you accept *or* reject your hypothesis? Why?
   II. Hypothesis 2: do you accept *or* reject your hypothesis? Why?
  III. Hypothesis 3: do you accept *or* reject your hypothesis? Why?
  IV. Hypothesis 4: do you accept *or* reject your hypothesis? Why?

4. OPTIONAL: REDUCING THE IMPACTS OF STORMWATER RUNOFF

The final step is to use what you have learned from your observations, measurements, and the assessment of your hypotheses to answer the more overarching question of what can be done to reduce the impact of stormwater runoff.

Below you will be listing commonly used best management practices (BMPs) that have been adopted for campuses. (**Best management practices (BMPs)** describe practices, planning, policies, and engineered devices and systems to prevent, reduce, control, or treat contaminated stormwater.) For each of the following, use the Internet to identify an example of an adopted BMP. Be sure to briefly describe the BMP, identify how it will positively impact stormwater, and properly cite your source:

  i. Reducing or eliminating impervious surfaces or the imperviousness of existing surfaces

 ii. Reducing generation of pollutants on surfaces

iii. Capturing stormwater

iv. Treating contaminated stormwater

This is a formal lab write-up. Be sure to use the following headings:

WRITE-UP

Title:

  I. **Introduction:** Discuss the generation and impact of stormwater based on peer-reviewed literature. Discuss the impacts of stormwater including contaminated stormwater on a campus. Be sure to end this section with your hypotheses.

 II. **Methods:** Discuss how you collected your data.

III. **Results:** Present your results including a results table and photographs as necessary. Be sure to include your answers to the questions on the amount of rainfall on your respective study areas, the percentage of runoff, pollutants, slope, and flow direction.

IV. **Discussion:** Discuss whether you accepted or rejected your hypotheses. Relate your findings to real-life applications such as the name, distance, and location of the nearest water body potentially impacted by stormwater runoff. Optional: in a separate paragraph, include your discussion of BMPs.

 V. **References Cited**

Applying the Scientific Method: Dowsing for Water

OBJECTIVES

- Apply the scientific method to design an experiment.

- Critically evaluate scientific evidence about a controversial environmental issue or technique.

KEY CONCEPTS AND TERMS

✓ Dowsing

TIME REQUIRED: 1 hour of Internet research, 2 hours to design and carry out an experiment.

INTRODUCTION

Dowsing is a method that uses a Y-shaped divining rod or two L-shaped rods to search for underground water, metals, or other valuable resources. It is an ancient practice, likely dating back 10,000 years. By the 15th century, German practitioners were specialized in finding precious metals with "divining rods." In recent years, dowsing has been used to search for buried water lines as well as potential locations for groundwater wells. In water-rich states like those of New England, it is quite common to encounter water. In the west and southwest, water sources are fewer and deeper so it can be more difficult and thus costly to pinpoint sources. Statistics and probability play a role in evaluating the likelihood of water availability, as do geological characteristics.

How does dowsing work? Some theories suggest that there is a psychic connection established between the dowser and the sought object because all living things supposedly possess an energy force. By concentrating on the hidden object, the dowser is able to tune into the energy force or "vibration" of the object, forcing the dowsing rod or stick to move. In this case, the divining tool may act as an amplifier or antenna for tuning into the energy. There is another view: dowsing is a myth based not on science but on superstition, and dowsers are successful in part because they "read" the landscape quite effectively and in part because groundwater is easily found in many places. In 1852, William Carpenter proposed that the rods move due to involuntary motor behavior, which he termed "ideomotor action" (Carroll, 2003). The Smithsonian Institute has reported the lack of scientific evidence on dowsing (Zielinski, 2009). Dowsing is controversial to the public and therefore presents an interesting example for the application of the scientific method.

Critical thinking is a key element in science. Like many other sciences, environmental science can quickly become controversial when it challenges core beliefs. Avoiding controversy can actually impede solving environmental problems. More and more, scientific associations and educators are promoting engagement in controversial questions, particularly where the public might be affected (e.g., SENCER, 2017). For this lab, you will be applying the scientific method to dowsing for determining some testable explanations. See Figure 15.1.

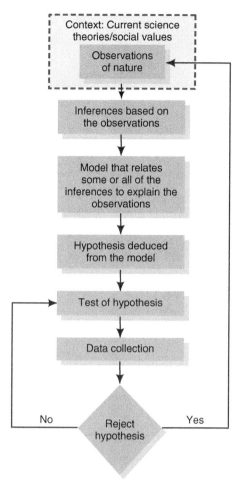

FIGURE 15.1 Scientific method

• Dowsing rod or stick

There are at least five variables or sets of variables (components or classes of variables) involved in dowsing: the *dowser, dowsing instrument, air, ground*, and *water* (or other material being sought) as shown in Figure 15.2. Can you think of any other classes of variables that might affect dowsing?

TASKS

With a partner or your lab group, complete the following tasks:

1. Find an article in the library or on an Internet site that is pro dowsing (a good place to start is the American Society of Dowsers, 2017) and an article or site skeptical or critical of dowsing (e.g., James Randi Educational Foundation, 2015). Summarize what each document or source says. Does it seem credible? How do you evaluate its credibility? What strengths and flaws do you detect? Be sure to cite the sources in your lab report.

2. Write an alternate hypothesis for one of the five categories of variable: dowser, dowsing instrument, air, ground, and water (Figure 15.2).

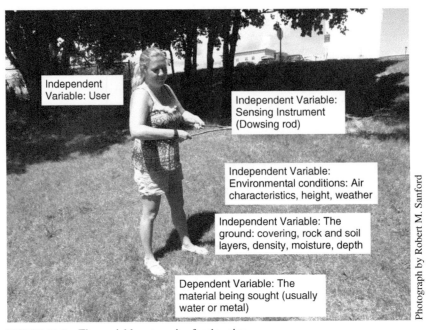

FIGURE 15.2 Five variable categories for dowsing.

The first four are independent variables: User characteristics, measuring instrument characteristics (dowsing rod), environmental conditions (including air, weather, distance from ground), and the medium that contains what you are seeking (ground cover, rock and soil layers, density, moisture, depth). The dependent variable is the water, metal, or other material being sought.

3. Suppose your hypothesis about the dowsing instrument is that dowsing works better for metal rods than rods made of other materials because metal rods "pick up" electromagnetic emanations from the water. You would want to have a reference in your introduction to support your decision to test this hypothesis; in the absence of references, you could present a reasoning for the test, such as "metal is a good conductor." You could test by controlling for everything except the type of material in the rod. You could use different types of metals and also glass, plastic, or wood. In this experiment, you might suggest blindfolding the testers so that they do not know which type of rod they are using.

   A. In the above example, how many testers would you suggest and why?

   B. How many trials (repeats) would you suggest for each material?

   Write a simple experimental design to test your hypothesis. Modify it as needed so that you can test it with the materials at hand. Be sure that the description of the experiment is sufficient to allow someone else to conduct the same experiment. Use a diagram if it helps. You want to have at least 20 trials or 20 test subjects.

4. Select one of the factors and conduct the experiment you described above to test your hypothesis.

   A. What were the results? Give them in a table in your results section, but save your analysis for your discussion section.

   B. Do you have enough data to determine whether or not the results support your hypothesis?

5. Did dowsing "work" for your subject (the person who did the dowsing)?

6. In your discussion section, analyze your results. What have you determined about the scientific basis? How confident are you in your results? Why? Be sure to tell how you would modify your experiment for further research. What future experiment can you conduct to increase your confidence?

7. Include a bibliography or references section. Be sure that you have proper reference citations.

8. How did this activity help you to think about the application of scientific reasoning to the environmental sciences? What do you think about science as a tool to address controversial issues?

For this lab, a formal experimental laboratory report is required:

WRITE-UP

Title:

I. **Introduction:** What did you do and why did you do it? Discuss the issue of dowsing and of groundwater availability in your area—this is your "literature review." If you use a website, be sure to check out some of the comments and discussion sections on whether or not dowsing "works." Be sure to have references for any information you cite or otherwise use in your report. State your hypotheses (e.g., your expectation of your experimental results), including the one you choose to test, which should be a logical outgrowth of your introduction and literature review.

II. **Methods:** How did you collect your data?

III. **Results:** What did you find? Be sure to include tables and graphs as appropriate.

IV. **Discussion**
   - Describe the potential areas of weaknesses in your conclusions (e.g., sampling, interfering variables, extrapolation, and assumptions).
   - Examine and reflect on this data as an individual. That is, what is the meaning of your findings? Were your samples large enough? Were you surprised by any of the data? What is your reasoned opinion about dowsing? Have you addressed Task items 1 through 8?

V. **References Cited**

REFERENCES

American Society of Dowsers. 2017. American Society of Dowsers, Home. Available at http://www.dowsers.org (verified 26 July 2017).

Carroll, R.T. 2003. The Skeptic's Dictionary. Wiley & Sons, New York.

James Randi Educational Foundation. 2015. James Randi Educational Foundation. Available at Web.randi.org (verified 26 July 2017).

SENCER. 2017. SENCER Science Education for New Civic Engagements and Responsibilities. Available at Sencer.net (verified 26 July 2017).

Zielinski, S. 2009. Debunking Dowsing. Smithsonian.com. Available at http://www.smithsonianmag.com/science-nature/debunking-dowsing-5028261 (verified 26 July 2017).

Aquatic Species Diversity and Water Quality

OBJECTIVES

- Identify and use biodiversity terminology and concepts.

- Sample invertebrate residents of a pond, lake, or stream to estimate and assess biodiversity.

- Relate biodiversity to water quality.

KEY CONCEPTS AND TERMS

✓ Biodiversity

✓ Ecological diversity

✓ Functional diversity

✓ Genetic diversity

✓ Macroinvertebrates

✓ Sequential Comparison Index (SCI)

✓ Species diversity

✓ Species evenness

✓ Species richness

✓ Water quality

TIME REQUIRED: 2 hours in addition to travel to and from sample site.

INTRODUCTION

Biodiversity is a popular term used by public officials, the media, the public, and environmentalists. What exactly does it mean? **Biodiversity** (or biological diversity) is defined as different life forms (species) and life-sustaining processes that can best survive the variety of conditions found on the earth. More specifically, biodiversity includes **genetic diversity** (variety in genetic makeup within a species), **species diversity** (variety among species in habitats), **ecological diversity** (variety of habitats such as forests, grasslands, and streams), and **functional diversity** (biological and chemical processes or functions). Biodiversity is a proxy indicator of environmental quality. Human actions, such as agriculture, elimination of species, pollution, deforestation, pesticide application, armed conflicts, and settlement, can reduce biodiversity. This reduction in biodiversity results in both known and unforeseen consequences as ecosystems react to these disturbances. Loss of biodiversity can increase the vulnerability of an ecosystem to further perturbations by reducing its stability and functionality.

This lab's focus is on species diversity defined as the numbers and proportions of different species found in an ecosystem. Knowing something about species diversity within a habitat provides information about the complexity of the ecosystem. Complexity sometimes correlates with stability because a complex system can be more resilient to stress. Environmental scientists use indices of diversity in conjunction with *rareness* (the spatially restricted occurrence of a species) and habitat evaluation to make inferences on the state of an ecosystem. There are two components to measure this diversity: species richness and species evenness.[1] **Species richness**

---

[1] Calculations of richness and evenness can get complicated for site comparisons. One example of free software available on the Internet to do this is EstimateS 9, developed by Dr. R. Colwell (2016) for Windows and Macintosh. It addresses biodiversity functions, estimators, and indices based on biotic sampling data. The software and a user's guide are available for free from the University of Connecticut at http://viceroy.eeb.uconn.edu/EstimateS.

is the number of different species. **Species evenness** is the relative proportion of each species. The **Sequential Comparison Index** (SCI) is a simple method used to evaluate biodiversity and environmental quality in aquatic ecosystems (Cairns et al., 1968). The index ranges from 0.0 to 1.0. Although this index is only a general approximation, it is useful for comparisons if properly done. A high index (0.6 or more) suggests high biodiversity, high **water quality**, and dominance of species intolerant to pollution. A low index (0.3 or below) indicates low biodiversity, low water quality, and species with high tolerance of pollutants.

MATERIALS

- Dip nets or kick nets
- Dissolved oxygen meter
- Forceps
- Macroinvertebrate key for local species
- Notebook
- pH meter
- Squeeze bottles
- Temperature probe
- Trays (white, metallic, or light color, plastic or metal)
- Waders, boots, or hip boots

TASKS

In this lab, you will gather and sort samples of **macroinvertebrates** (visible organism without a backbone, often an insect) from a local pond, lake, or stream to determine biodiversity. You will also take some basic water quality measurements.

1. In teams, use a net to take a sample of macroinvertebrates in the water from the edge of a pond/lake/stream. In addition, if you are measuring water quality at each sample site, measure the temperature, dissolved oxygen, and pH at each sample site. Employ the following steps for collecting macroinvertebrates with a dip net:
   - Choose a suitable habitat (leaf packs, snag areas, banks, and/or sediment) to collect samples.
   - When collecting from a stream, be sure to move from downstream to upstream so the current keeps the invertebrates in the net. For all habitats, be sure you do not disturb the site before sampling.
   - If collecting from a leaf pack, very gently drag the dip net through the leaf pack, collecting leaves and debris.
   - To collect from a snag area or a bank, gently drag across the snag or the bank with the net to collect debris.

2. Dump the net's contents into a tray. Carefully remove the debris from any living organisms. Use a squeeze bottle filled with pond water to gently wash away debris from macroinvertebrates. Some aquatic insects build cases and cocoons, so be very careful in differentiating between debris and a macroinvertebrate.
   - Try to keep or align the specimens into rows in the same relative positions that they landed in the tray. Keep specimens wet.
   - The specimens may move around quite a bit, but this is the nature of field science, so do the best you can.

3. Once you have at least 12 organisms in a sample, begin in the upper-left corner of the tray and count horizontally, scoring the number of adjacent organisms that are of the same type. Each group of one or more organisms is called a **run**. A run ends when you encounter a different type of organism. At the end of the row, continue into the next row as if there was no break (i.e., this is a continuous process). Thus, if the last organism in a row is the same organism type as the first one in the next row, it is the same run. If you are unable to get 12 individuals despite repeated sampling, move to another location or pool your samples with other investigators.

4. Each individual organism will be noted with a letter. When you encounter a different species, this represents new run. Assign it a new letter in sequence (e.g., a,a,a,b,b,a,c,c,b,b,a,a = 12 individuals in 6 runs). Now, divide the number of runs by the number of individuals to obtain SCI (= Runs/Individuals). In this example, 6/12 = 0.5, therefore SCI = 0.5. The SCI will range from 0 to 1. A low SCI number (0–0.3) means low diversity, whereas an SCI of 0.7 to 1 indicates high diversity. Figure 16.1 provides another example of how to determine an SCI.

5. Record the SCI into Table 16.1 and provide a brief description of the habitat you sampled and describe the characteristics of each sampling site, which should include degree of sunlight, slope, bottom conditions, vegetation, cover, etc.

6. Repeat this process to collect five complete samples sufficient to calculate SCIs. Choose different areas in the pond/stream to sample. If you do not get enough data to compare between your samples, sum up your trials into one sample and compare with the results of other students.

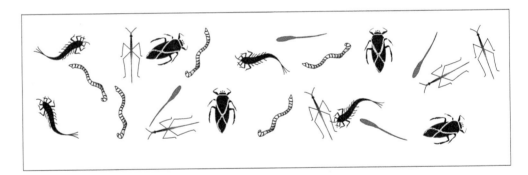

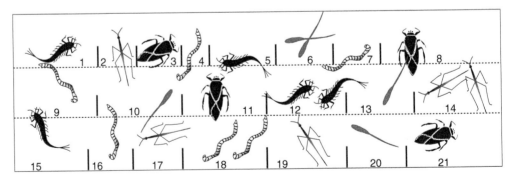

$$SCI = \frac{\text{No. of runs}}{\text{No. of individuals}} = \frac{21}{25} = 0.8 = \text{Very diverse}$$

FIGURE 16.1 Determining a sequential comparison index (SCI)

The SCI provides an easy way to obtain an approximation of species diversity and ecosystem health. Ideally, it should be used in conjunction with other tests.

Table 16.1   SCI Data Collection Sheet

Sample number	SCI	Habitat description
1		
2		
3		
4		
5		

7. Use an existing map of the waterbody (or draw your own using Google Earth as a guide) and clearly mark your sampling areas.

Your lab report will be on biodiversity and water quality for a waterbody and should contain the following:

WRITE-UP

Title

I. **Introduction** (the name, location, and characteristics of the waterbody)

II. **Methods** (describe your sampling procedure)

III. **Results** (be sure you address the following items)
   - For each sample site, what was the SCI and description of the basic habitat?
   - Using your data, create a graph that depicts the SCI for multiple sample locations. Report the mean SCI for the waterbody.
   - Report your DO, pH, and temperature measurements.

IV. **Discussion**
   - Comment on the species richness and species evenness.
   - Comment on the water quality of your waterbody. What other data would you want to collect to determine the ecological and environmental health of the pond/stream?
   - What human activities might have adversely affected species diversity in this pond/stream? Are there any particular vulnerabilities you see on your site that could affect water quality and biodiversity? For example, is there a nearby parking lot that could contribute runoff?
   - How has this lab activity influenced your thoughts about biodiversity and water quality of the waterbody?

V. **References Cited**

REFERENCES

Cairns, J., D.W. Albaugh, F. Busey, and M.D. Chaney. 1968. The Sequential Comparison Index—A simplified method for non-biologists to estimate relative differences in biological diversity in stream pollution studies. *Journal of the Water Pollution Control Federation.* 40:1607–1613.

Colwell, R.K. 2016. *EstimateS*, Version 9: Statistical Estimation of Species Richness and Shared Species from Samples (Software and User's Guide). Freeware for Windows and Mac OS. Available at http://viceroy.eeb.uconn.edu/EstimateS (verified 20 July 2017).

Environmental Forensics

OBJECTIVES

- Compete with several field and lab techniques to detect environmental contaminants.

- Identify the potential cause and source of a reported fish kill using an analytical and deductive investigative approach.

KEY CONCEPTS AND TERMS

✓ Ammonia

✓ Biological oxygen demand

✓ Dissolved oxygen

✓ pH

✓ Point source discharge

✓ Specific conductance

✓ Turbidity

TIME REQUIRED: Prelab task 1 hour and main lab 1.5 hours.

INTRODUCTION

In the environmental field, we often observe environmental conditions and problems without immediately knowing their root cause. Some of them are chronic problems, such as frog deformations, species decline, and global climate change, and others are acute problems, including beaching of whales, viral infections of deer, and pesticide poisonings. Environmental scientists are often called into an investigation to determine what happened, how to remediate the problem, and how to prevent future problems. This backtracking, investigative approach is sometimes referred to as environmental forensics, requiring the skills of an environmental detective.

The approach is to use all available evidence to find the likely cause. Chemical and physical analyses are crucial, but they are not the only methods available to identify the root cause of a problem. Deductive reasoning by using anecdotal evidence, experience, field observations, and previous studies and reports is an essential component in solving the mystery.

BACKGROUND

On August 26, a massive fish kill occurred in the Mountain River approximately 1 km downriver from Mountain City. Various local residents reported the kill to the state Bureau of Environmental Protection on the evening of the 26th. Some residents claimed that they saw hundreds of Rainbow trout (*Salmo gairdneri*) floating on their sides and bellies. According to one report, some of the trout were gasping. There were conflicting reports about visual water quality. Some reports mentioned that the water was darker than normal. Also, some reports stated that there was some type of "foam" on the water. No oil sheens were reported. No reports of spill of hazardous materials were made to the state police or local fire department.

A local resident recorded the river temperature where the largest concentration of dead fish was observed, which was 26°C. The measurement was taken just below the surface. On August 26, the weather was reported as cloudy and overcast, with an ambient temperature of 30.5°C. The 8 days prior to the 26th saw unusually high ambient temperatures, no precipitation, and uninterrupted overcast skies. At the nearby airport, the winds on the 26th were reported as relatively calm—7 km/hour from the north.

The Mountain River watershed is approximately 125 km². Except for Mountain City, there is relatively sparse human population in the watershed. The housing within the river's watershed is primarily vacation homes and hunting camps. As shown in Figure 17.1, there are three facilities permitted by the state to discharge effluent into the river through point sources: the Mountain City sewage treatment plant, Riverside Brewery, and ABC Chemical Company. (A **point source discharge** is a discrete, fixed discharge point such as a pipe, culvert, ditch, or tunnel.) All three facilities are located in Mountain City; they range from 0.9 to 1.8 km upstream from the reported fish kill. Note that with point source discharges, because of the physical characteristics of rivers, there are varying zones of pollution impact as shown in Figure 17.2.

The Mountain City sewage treatment plant is relatively small. It is 26 years old and serves approximately 1,250 residents and 8 industrial dischargers. The plant employs primary and secondary treatments and uses chlorine as a disinfectant prior to discharge into Mountain River. The Riverside Brewery is located across from the sewage treatment plant, but is not connected to the sewage plant and discharges directly into Mountain River. Approximately 0.9 km upstream from these two facilities is the ABC Chemical Company. This company does not manufacture chemicals, but purchases sulfuric and phosphoric acids in bulk and repackages them for the consumer market.

Approximately 25 years ago, the state's Fish and Wildlife Department began stocking Rainbow trout (*S. gairdneri*) in Mountain River. The fishery has become well established and is recognized for producing trophy trout.

In general, trout tend to be a more sensitive species, environmentally speaking. Like all species, *S. gairdneri* has a range of tolerance for environmental conditions (the concept of tolerance is explored in Lab 7; see Figure 7.1). Although they prefer cold, well-oxygenated waters, Rainbow trout are capable of surviving in waters as warm as 29.4°C (85°F) provided that the

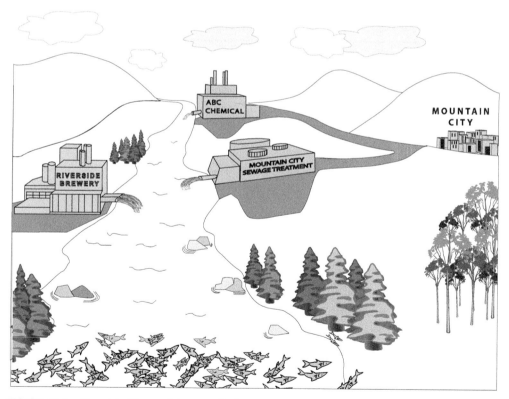

FIGURE 17.1   Mountain River point source discharges

This graphic represents the study area and the general location of the different discharges, which intermingle as the water flows downstream (not to scale).

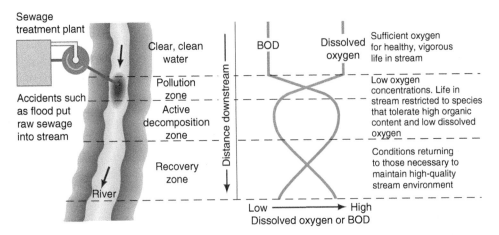

FIGURE 17.2 Sample pollution impact zones in a river

The input of sewage affects the relationship between dissolved oxygen and **biological oxygen demand** (BOD). Dead organic matter produces BOD, which can cause the amount of dissolved oxygen to become too low to sustain life.

Source: Botkin, D.B., and E.A. Keller. 2007. Environmental Science: Earth as a Living Planet, 6th ed. John Wiley & Sons, New York (p. 467).

water remains well aerated. Rainbow trout generally feed off the bottom, foraging on virtually all aquatic insects and their larvae.

States and most tribal lands have their own water quality standards, which are based on the U.S. Environmental Protection Agency's National Recommended Water Quality Criteria. The following are common parameters of surface water quality potentially relevant to fish kill:

- Specific conductance

- Dissolved oxygen

- Turbidity

- Ammonia

- pH

SPECIFIC CONDUCTANCE

Specific conductance (SC) (or conductivity) is a measure of the ability of water to pass an electrical current and is an indicator of potential water pollution. Conductivity of water is affected by the presence of inorganic dissolved solids such as chloride, nitrate, sulfate, and phosphate anions (ions that carry a negative charge) or sodium, magnesium, calcium, iron, and aluminum cations (ions that carry a positive charge). Organic compounds such as oil, phenol, alcohol, and sugar do not conduct electrical current very well and therefore have low conductivity in water (US EPA, 1997). Conductivity is also affected by temperature: the warmer the water, the higher the conductivity. Because conductivity readings can fluctuate with temperature, a standardized measurement called specific conductance (SC) is often used. Because conductivity readings can fluctuate with temperature, a standardized measurement of specific conductance (SC) is made at 25°C and thus is reported as conductivity at 25°C ; it can be calculated from known conductivity and temperature values as follows:

$$SC = \frac{Conductivity}{1 + 0.0191(t - 25)}$$

where t = temperature of sample.

The basic unit of measurement of conductivity is siemens.[1] Conductivity is measured in microsiemens per centimeter (μS/cm). Distilled water has conductivity in the range of 0.5 to 3 μS/cm. The conductivity of rivers in the United States generally ranges from 50 to 1500 μS/cm. Studies of inland fresh waters indicate that conductivity of streams supporting good mixed fisheries ranges from 150 to 500 μS/cm. Conductivity outside of this range could indicate that the water is not suitable for certain species of fish or macroinvertebrates. Industrial waters can exhibit as high as 10,000 μS/cm (US EPA, 1997).

Dissolved oxygen (DO) is a measure of the amount of oxygen freely available in water. It is commonly expressed as a concentration in terms of milligrams per liter (mg/L) or as percent saturation, which is temperature dependent. Percent saturation is the percentage of the potential capacity of water to hold the oxygen that is present. DO levels fluctuate seasonally and over a 24-hour period. They vary with water temperature and altitude. Cold water holds more oxygen than does warm water and water holds less oxygen at higher altitudes (US EPA, 1997). Thermal discharges, such as water used to cool machinery in a manufacturing plant or a power plant, raise the temperature of water and lower its oxygen content. Aquatic animals are most vulnerable to lowered DO levels in the early morning on hot summer days when stream flows are low, water temperatures are high, and aquatic plants have not been producing oxygen since sunset (US EPA, 1997).	**DISSOLVED OXYGEN**

The DO for surface water ranges from 0 mg/L in extremely poor water conditions to a high of 15 mg/L (15 ppm) in 0°C (freezing) water. The optimal DO level for *S. gairdneri* is 7 to 9 mg/L, and the acute lethal limit for trout is 3 mg/L (US EPA, 1986).

Turbidity is the measurement of water clarity and is related to the amount of suspended solids in water. Suspended solids range from clay, silt, and plankton to industrial wastes and sewage. Turbid waters become warmer as suspended particles absorb heat from sunlight, causing DO to decrease. Suspended solids also can clog fish gills. Turbidity is measured in nephelometric turbidity units (NTUs). A normal range for turbidity in river water has not been established. Turbidity in drinking water should be less than 1 NTU. The US EPA standard for drinking water is 0.5 to 1.0 NTU (US EPA, 1986).	**TURBIDITY**

Sewage is the major source of anthropogenic **ammonia** (NH_3) in surface waters. Ammonia comes from the decomposition of urea in urine and the decomposition of nitrogenous materials in sewage. Ammonia is toxic to fish and aquatic organisms, even in very low concentrations. When levels reach 0.06 mg/L, fish can suffer gill damage. Levels of 0.2 mg/L cause sensitive fish such as trout and salmon to die (US EPA, 1986).	**AMMONIA**

pH measures the acidity or alkalinity of a solution (Figure 17.3). Acid conditions are usually caused when there is an excess of hydrogen ions (H^+). Alkaline conditions occur when there is an excess of hydroxyl ions (OH^-). At pH 7 (known as neutral), there is an exact balance between OH^- and H^+ ions; thus, it is neither acidic nor alkaline. As shown in Figure 17.3, the pH scale is logarithmic, which means that there is a tenfold change in the number of H^+ ions (acid-causing particles) or OH^- ions (alkaline-causing particles) present for each change in pH of 1 unit. Thus, pH 5 is 10 times more acidic than pH 6 and 100 times more acidic than pH 7. pH affects many chemical and biological processes in water. For example, different organisms flourish within	**pH**

---

[1] The SC instrument measures electrical resistance, that is, the extent that something can *resist* an electrical current, which normally is reported in ohms. The unit of conductance was originally a "mho" (*ohm* spelled backward). More recently, the term **siemens** is used in accordance with the terminology of the International System of Units. Both mho and siemens are sometimes used in water quality reports. 1 siemen is equal to 1 mho. Because SC in natural waters is usually much less than 1 siemen/cm, SC is usually reported in microsiemens (1/1,000,000 siemens) per centimeter, or μS/cm.

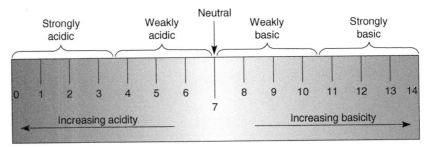

FIGURE 17.3 The pH scale

Pure water is neutral with a pH of 7. Each decrease of 1 pH unit means that the acidity has increased tenfold.

Source: Raven, P.H., L.R. Berg, and D.M. Hassenzahl. 2008. Environment, 6th ed. John Wiley & Sons, New York (p. 471).

different ranges of pH. A greatest variety of aquatic animals prefer a pH of between 6.5 and 8.0. A pH value outside of this range reduces the diversity in the stream because it stresses the physiological systems of most organisms and can affect reproduction. In addition, a low pH value can allow toxic elements and compounds to become mobile and available for uptake by aquatic plants and animals, resulting in secondary toxic effects (US EPA, 1997).

MATERIALS

There are four containers holding samples prepared by the instructor that represent samples taken from the river at four specific locations:

A. Upstream from Mountain City (this is your background sample, or control)

B. Immediately downstream from the Mountain City sewage treatment plant

C. Immediately downstream from the Riverside Brewery

D. Immediately downstream from the ABC Chemical Company

You will be provided with various chemical, physical, and analytical testers; test kits; and reference materials, which may include the following:

- Ammonia test
- Conductivity meter
- Dissolved oxygen meter
- pH meter
- Thermometer
- Turbidity meter
- Other tests/meters

TASKS

1. As a team, read the background material carefully. Based on the background information, make an initial hypothesis as to the primary cause of the fish kill, which can include one of the four samples, weather, nature, etc.

2. Your job is to test and analyze *every* aspect of the water with the equipment provided to identify the primary root cause of the fish kill. (Use Table 17.1 to input your results and be sure to specify your units.)

Table 17.1 Data Collection Sheet for Spill Investigation

Parameter	Background	Riverside brewery	ABC chemical company	Sewage treatment plant
Turbidity				
NH_3				
DO (ppm)				
pH				
Conductivity				
Other				
Other				
Other				

3. Conduct some basic research to compare your findings with the references. You can check with your state or consult the U.S. Environmental Protection Agency's National Recommended Water Quality Criteria – Aquatic Life Criteria Table (https://www.epa.gov/wqc).

4. Employ logic and deductive reasoning in an attempt to pinpoint the likely source based on your knowledge of the source's potential impacts on water quality, the impact of the water quality on trout, and the results of your water quality analysis.

Your lab report will be a memorandum to the director of the Bureau of Environmental Protection. Use the following format:

WRITE-UP

MEMORANDUM

TO:

FROM:

DATE:

SUBJECT:

Paragraph 1—Background explaining the problem, 2 or 3 sentences

Paragraph 2—The likely water quality impacts of each potential source

Paragraph 3—What you tested for, a summary of your results, any uncertainties

Paragraph 4—Your conclusion as to the primary root cause of the fish kill

Paragraph 5—Your recommendations to the director

Attachment: A table presenting your analytical results

NOTE: The results of your investigation are likely to be the basis of an enforcement action and will become public record. Therefore, it is imperative that your memorandum be written in neutral language, presenting an objective analysis of the facts. Avoid inflammatory and value-laden terms and opinions.

U.S. Environmental Protection Agency (US EPA). 1986. Quality Criteria for Water. Washington, DC.

U.S. Environmental Protection Agency (US EPA). 1997. Volunteer Stream Monitoring: A Methods Manual [Online]. Available at http://www.epa.gov/volunteer/stream/index.html (verified 17 June 2008).

REFERENCES

Actual/Virtual Field Trip: Municipal Wastewater Treatment Plant

OBJECTIVES

- Identify the location of a local municipal wastewater treatment facility.

- Describe the basic processes in treating sewage, industrial wastewater, and stormwater runoff.

- Discuss the current and future challenges facing municipal wastewater treatment facilities.

KEY TERMS AND CONCEPTS

✓ Anaerobic digestion

✓ Biosolids

✓ Combined sewer overflow (CSO)

✓ Effluent

✓ Municipal wastewater treatment facilities (WWTF)

✓ Stormwater runoff

TIME REQUIRED: 2 to 3 hours for actual field trip and 1 to 2 hours for virtual field trip.

INTRODUCTION

Most **municipal wastewater treatment facilities (WWTFs)** were built in the last quarter of the 20th century to comply with federal legislation. Before that, most municipal systems simply dumped raw sewage into public waters. During the 1970s and 1980s, under the Clean Water Act, the US EPA funded more than $60 billion for the construction of publicly owned and operated wastewater treatment projects (US EPA, 1998). These projects included sewage treatment plants, pumping stations, and the rehabilitation of aging sewer systems. This program led to the dramatic improvement of water quality in thousands of municipalities nationwide.

The purpose of a WWTF is to treat sewage, industrial wastewater, and often some stormwater runoff. **Stormwater runoff** is generated from precipitation snowmelt events that flow over impervious surfaces, such as paved streets, sidewalks, parking lots, and roofs, or land when it infiltrate into the ground. The typical modern sewage treatment plant employs primary and secondary treatment (Figure 18.1). WWTFs utilize bacteria to emulate and accelerate the natural process of organic degradation. Primary treatment involves screening to remove debris (wood, plastic, litter, etc.) and grit. The wastewater is also aerated to increase the oxygen level. Secondary treatment breaks down the organic matter using microbes in combination with oxygen. Oxygen is essential because decomposition of organic matter consumes oxygen thereby requiring replenishment. Aeration also further enhances the removal of grit. The next step is the separation process where water is separated from solids (sludge). The water part is often further treated by filtration to remove bacteria and other fine materials. The sludge, called **biosolids**, is

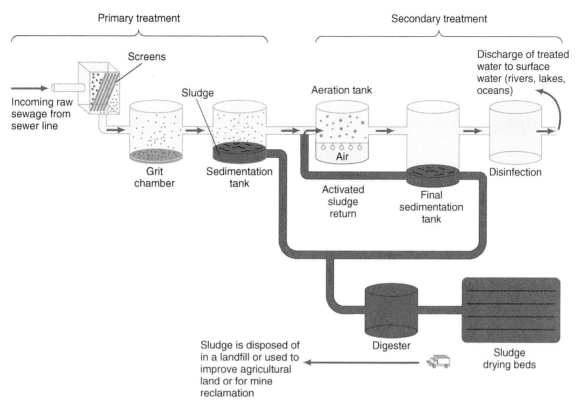

FIGURE 18.1   A municipal sewage treatment plant

Source: Botkin, D.B., and E.A. Keller. 2007. Environmental Science: Earth as a Living Planet. 6th ed. John Wiley & Sons, New York (p. 481).

removed where it may be treated on-site through **anaerobic digestion** (use of microorganisms to break down biodegradable material in the absence of oxygen), which reduces the volume, eliminates odors, and kills disease-causing organisms. A byproduct of anaerobic digestion is methane gas, which can be captured and used as a fuel. The finished product from the digestion process is generally sent to landfills but sometimes can be used as fertilizer depending on the presence of harmful constituents. The final stage of the water, the **effluent** (liquid discharged into a receiving water), is disinfected using chlorine, ozone, or UV light before discharge to reduce harmful bacteria.

However, WWTFs are not limited to treating sewage. Many industries may discharge pre-treated wastewater to WWTFs. In addition, many WWTFs also treat collected stormwater. One particular design, the **combined sewer overflow (CSO)** system, is designed to collect stormwater, sewage, and industrial wastewater in the same pipe. Under normal conditions, such a system is able to properly treat all three inputs; however, during heavy precipitation events, these systems can exceed the design capacity, meaning that untreated sewage, stormwater, and industrial wastewater are discharged directly into the receiving water.

In this lab, you will be going on an actual or virtual field trip of a WWTF. During either type of field trip, seek out answers to the following questions to help you prepare your write-up in order to achieve the lab's learning objectives.

TASKS

QUESTIONS TO BE ADDRESSED AFTER THE FIELD TRIP

1. What is the name of the facility you visited? (If it was a virtual tour, provide the URL for the tour.)

2. Where is the WWTF located?

3. What is the geographic area it serves?

4. How many people does it serve? What is the status of the area population since the construction of the facility (i.e., increased or decreased, and to what percent)?

5. When was the facility built?

6. What is its design capacity?

7. How many industries and what type of industries are connected to the system?

8. Does the facility receive stormwater runoff for treatment?

9. Is the facility a CSO system? If yes, what challenges have faced the facility by being a CSO system?

10. How does the WWTF remove solids, reduce organic matter, and disinfect the effluent?

11. What types of parameters or conditions are monitored by the facility operators while the wastewater is being processed?

12. Where does the facility's treatment sludge (biosolids) go? Who monitors the management of the sludge to ensure that it meets health and environmental standards? What are some of the standards? What happens if the sludge fails to meet the standards?

13. How can sludge become a useful byproduct of the WWTF system?

14. Where is the treated water discharged (i.e., name of the receiving surface waterbody)?

15. Who specifically monitors the discharge of treated effluent to ensure that it meets the standards? What are some of the standards? What happens when they are not met?

16. If the water is safe to be discharged into the environment, why is it not used as the drinking water source for the community?

17. What seasonal changes (if any) are necessary in the operation of the facility?

18. What is the biggest water quality concern facing the facility? What plans, if any, do the operators have for addressing the concern?

19. As water users and waste generators, we all contribute to the wastewater stream. (Even if you have an on-site private septic system in the service area, some waste eventually comes to the facility, and you may even have to pay an "availability fee" if you are near but not connected to a municipal system.) What specific types of pollution prevention actions can be undertaken to reduce the volume, pollutant loads, and toxicity of wastewater sent to the facility?

WRITE-UP

The write-up for this lab is a technical report. Use the following headings for your report. The tone of your report should be neutral and fact-based.

Title:

    I. Introduction (name and location of the WWTF, geographic area served by the WWTF, year built, capacity, and date and time of field or virtual site visit and the URL for virtual visits)

II. Operation of the WWTF (describe the basic processes in treating sewage, industrial wastewater, and stormwater runoff)

III. WWTF Operational Challenges (describe the current and future challenges facing the WWTF)

IV. References Cited

U.S. Environmental Protection Agency (US EPA). 1998. Wastewater Primer (EPA 833-K-98-001). Government Printing Office, Washington, DC.

REFERENCE

Actual/Virtual Field Trip: Wetlands and Their Ecosystem Functions

OBJECTIVES

- Describe the basic ecosystem functions and importance of wetlands.

- Explain the challenges faced in protecting wetlands.

- Report on the ecosystem functions of a specific wetland.

KEY TERMS AND CONCEPTS

✓ Ecosystem

✓ Ecosystem function

✓ Wetland

TIME REQUIRED: 3 hours for field trip option and 2 hours for virtual field trip option.

INTRODUCTION

Wetlands are an important type of ecosystem. (An **ecosystem** consists of plant, animal, and microorganism communities and their nonliving environment, interacting as a functional unit.) A **wetland** is defined, by sec. 404 of the Clean Water Act, as areas "that are inundated or saturated by surface water or groundwater at a frequency and duration sufficient to support, and that under normal circumstances to support, a prevalence of vegetation typically adapted for life in saturated soil conditions." There are many different types of wetlands including swamps, marshes, bogs, fens, prairie potholes, bayous, and others that differ in flooding regimes, dominant plant type and water characteristics such as pH or nutrient levels (Figure 19.1).

Wetlands perform many important **ecosystem functions** including water quality improvement, filtration and recharge of groundwater, floodwater storage, critical habitat for fish and wildlife, and biological productivity (see Figure 19.2 for example; there are many other ways to classify and evaluate functions and values). Wetlands are even artificially constructed to help treat stormwater and/or wastewater prior to its discharge or to perform other functions such as temporary storage of flood waters. These functions contribute direct and indirect values to society. For example, we can estimate the beneficial value of flood protection, water quality improvement, and the sale of hunting and fishing licenses as they relate to wetland-based habitat. In the intervening centuries since the United States became a country, over half of the nation's wetland areas have been lost or converted into other uses. Activities resulting in loss of wetlands and degradation include agricultural runoff and conversion into agricultural land, commercial and residential development, road construction, impoundment, mining, oil and gas development, industrial processes, creation of solid waste landfills, dredging, silviculture, and insect control (US EPA, 2017). Pollutants such as sediment, nutrients, pesticides, salts, heavy metals, acids, and selenium can degrade a wetland (US EPA, 2017). For this lab, you will visit (actual or virtual visit) a wetland to identify and assess its primary ecosystem functions and their potential value to society.

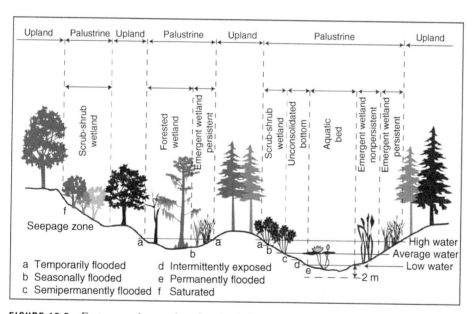

FIGURE 19.1   Photograph of a wetland

FIGURE 19.2   Features and examples of wetland classes and hydrologic modifications in the palustrine system

A simplified representation of a section across a bottomland hardwood forest from stream to upland, showing how various functions of interest to humans change across the transect. In actuality, the area is a complex spatial pattern of intermixed zones, not a linear gradient. This is a multivalue resource; different ecological processes peak in different zones, and these peaks are not directly related to their value to humans.

Source: Mitsch, W.J., and J.G. Gosselink. 2015. Wetlands, 5th ed. Wiley & Sons, New York.

<table>
<tr><td>

QUESTIONS TO BE ADDRESSED FOR THIS FIELD TRIP

</td><td>

1. What are the name, location, and size of the wetland? (Google Earth can be used.) If it was a virtual tour, provide the URL for the tour.

2. Is there an inlet or outlet to the wetland and if so, what are their names?

3. Who owns the wetland?

4. What type of wetland is it (swamp, marsh, bog, fen, bayou, etc.)?

5. List the primary ecosystem functions (flood control, wildlife habitat, water filtration, etc.) of this wetland and then rate each function (e.g., 1 to 5) in terms of its relative importance to the surrounding environment.

6. What species of ecological importance exist at the wetland?

7. What species of recreational value exist at the wetland?

8. What species of commercial value exist at the wetland?

9. Using the U.S. Fish and Wildlife Service's National Wetland Mapper (https://www.fws.gov/wetlands/Data/Mapper.html) and Google Earth, identify and describe the potential issues that currently affect the wetland or may impact the wetland in the future (e.g., population encroachment, development, roads, and so forth).

</td></tr>
</table>

WRITE-UP

The write-up for this lab is a technical report. Use the following headings for your report. The tone of your report should be neutral and fact-based.

Title:

   I. Introduction (name and location of the wetland, size of wetland, important historical or cultural features, and the URL for the virtual site visit)

   II. Type of Wetland (describe the wetland type, its prominent ecosystem functions, examples of important habitats, presence of unique or threatened/endangered species, and recreational opportunities)

   III. Wetland Status (describe current and potential future pressures or challenges facing the wetland and any discussion or plans to mitigate the potential impacts)

   IV. References Cited

REFERENCE

U.S. Environmental Protection Agency (US EPA). 2017. Wetlands Protection and Restoration. [Online]. Available at http://www.epa.gov/owow/wetlands/vital/status.html (verified 26 July 2017).

Actual/Virtual Field Trip:
Water Treatment Plant

OBJECTIVES

- Describe the process by which a public water district obtains, protects, treats, and distributes drinking water.

- Describe key issues associated with maintaining a safe drinking water system.

KEY TERMS AND CONCEPTS

✓ Potable

✓ Public water system

✓ Wellhead protection zones

TIME REQUIRED: 3 hours for field trip option and 2 hours for virtual field trip option.

Water districts supply water to large areas, treating water before it is delivered to the user as **potable** (drinkable) water and treating used water which then becomes sewage (Figure 20.1). The Safe Drinking Water Act (SDWA) is the federal law that protects public drinking water systems in the United States. A **public water system** is one that has at least 15 service connections or serves at least 25 people per day for 60 days of the year. In the United States, there are some 54,000 public water systems that provide drinking water to municipalities, condominiums, apartments,

INTRODUCTION

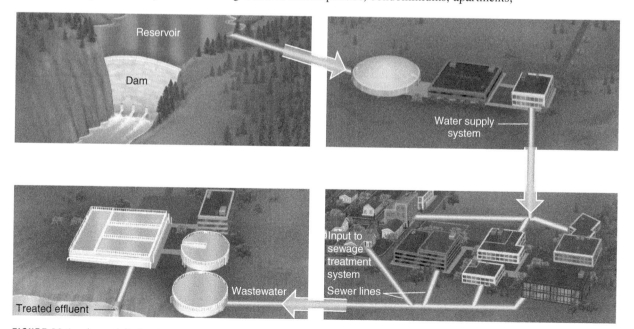

FIGURE 20.1 A municipal water treatment system

Source: Raven, P.H., L.R. Berg, and D.M. Hassenzahl. 2010. Environment, 7th ed. John Wiley & Sons, New York (p. 517).

and mobile home parks. There are other types of public water systems including transient systems, which may provide drinking water to a campground or rest areas, and nontransient systems, which supply water for a partial year, for example, schools.

Under the SDWA, the US EPA sets regulatory standards for drinking water quality, which currently covers 90 contaminants (see Figure 20.2 for example). The SDWA does not, however, regulate noncommunity private drinking water systems including private wells used by individuals. Public water systems must employ some level of treatment to destroy possible disease-causing microorganisms such as bacteria and protozoa and to remove a variety of other potential contaminants, natural or anthropogenic. Common treatment techniques include coagulation and flocculation, sedimentation, filtration, and, finally, disinfection. Disinfection historically used chlorine, but safer alternatives are increasingly being used such as ozone and UV light radiation. Obviously, the cleaner the water source, the less treatment is required. Thus, public water districts often adopt measures to protect watersheds when they are the source of potable water, which generally include use restrictions (e.g., no swimming, motorized boating, agriculture, etc.). Many public water systems use groundwater for their source. Similar to watersheds, **wellhead protection zones** (the surface and subsurface area surrounding a well used for drinking water) are established to prevent contamination of the well, which include restrictions on the use and disposal of chemical substances or installation of underground petroleum tanks within the wellhead protection zone.

NATIONAL DRINKING WATER STANDARDS	
CONTAMINANT	MAXIMUM CONTAMINANT LEVEL (mg/L)
Inorganics	
Arsenic	0.05
Cadmium	0.01
Lead	0.015 action level[a]
Mercury	0.002
Selenium	0.01
Organic chemicals	
Pesticides	
Endrin	0.0002
Lindane	0.004
Methoxychlor	0.1
Herbicides	
2,4-D	0.1
2,4-S-TP	0.01
Silvex	0.01
Volatile organic chemicals	
Benzene	0.005
Carbon tetrachloride	0.005
Trichloroethylene	0.005
Vinyl chloride	0.002
Microbiological organisms	
Fecal coliform bacteria	1 cell/100 ml

[a]Action level is related to the treatment of water to reduce lead to a safe level. There is no maximum contaminant level for lead.

Source: U.S. Environmental Protection Agency.

FIGURE 20.2 National drinking water standards

Source: Botkin, D.B., and E.A. Keller. 2014. Environmental Science: Earth as a Living Planet, 9th ed. John Wiley & Sons, Inc., New York.

In this actual or virtual field trip, you will explore where your drinking water source originates, how it is treated and distributed, and how the resource is protected. Obviously, the cleaner the water source, the less treatment it requires. Thus, we should be especially concerned about the activities occurring in the watershed (or wellhead zone) when it is the source of potable water.

Before you go on your virtual or actual site visit, use the US EPA's "Safe Drinking Water Information System" at https://www3.epa.gov/enviro/facts/sdwis/search.html. Click on your state and then find the drinking water plant that you plan to visit. The information will help you answer some of the following questions:

1. What is the name of the facility you visited? (If it was a virtual tour, provide the URL for the tour. Include the date of virtual or actual visit.)

2. Where is the drinking water plant located?

3. What is the geographic area it serves?

4. When was the plant built?

5. What is its design capacity?

6. How many people does it serve? What is the status of the area population since the construction of the facility (i.e., increased or decreased, and to what percent)?

7. How much water per day does it treat?

8. What is the source of the public plant's water? If the source is a public or private set of wells, where are they located and how are they protected? If the source is surface water, how is it protected?

9. Who are its major customer types (e.g., residential, commercial, schools, breweries)?

10. Is the demand for water expected to change due to population change or other reasons? If so, what are the plans to address this increased demand?

11. What are the plans for reducing water demands in the event of a short-term emergency (e.g., contamination)?

12. What are the plans for dealing with long-term problems such as drought?

13. What type of treatment is employed to make the water potable?

14. What type of energy is used to treat water at this plant? How is this energy supplied?

15. Has the water system had any violations? Use the US EPA's "Safe Drinking Water Information System" at https://www3.epa.gov/enviro/facts/sdwis/search.html. What are the major water quality concerns facing the watershed?

The write-up for this lab is a technical report. Use the following headings for your report. The tone of your report should be neutral and fact-based.

Title:

I. **Introduction**   (name and location of the water treatment plant, geographic area served by the plant, year built, capacity, and date and time of field or virtual site visit and the URL for virtual visits)

II. **Operation of the Water Treatment Plant**   (describe the basic processes in protecting the water source and treating and delivering clean, safe drinking water)

III. **Operational Challenges**   (describe the current and future challenges facing the plant such as future capacity concerns, potential impacts from climate change, and potential threats to drinking water from terrorism)

IV. **References Cited**

REFERENCE

Environmental Protection Agency (EPA). 2017. Analyze Trends: Drinking Water Dashboard [Online]. Available at https://echo.epa.gov/trends/comparative-maps-dashboards/drinking-water-dashboard (verified 26 July 2017).

Soil Characterization

OBJECTIVES

- Describe the process of digging a soil test pit.

- Identify soil horizons in the stratigraphy of a soil test pit.

- Describe soil characteristics using a color and texture guide.

- Identify the components of a soil report.

KEY CONCEPTS AND TERMS

✓ Parent material

✓ Pedosphere

✓ Soil horizon

✓ Soil profile

✓ Soil textural triangle

TIME REQUIRED: 2 to 3 hours.

Soil is a thin layer—called the **pedosphere**—on top of most of Earth's land surfaces. This thin layer is a precious natural resource. Because soils so deeply affect every other part of the ecosystem, they are often called the "great integrator." Soils hold nutrients and water for plants and animals. Water is filtered and cleansed as it flows through soils. Soils affect the chemistry of water and the amount of water that returns to the atmosphere to form precipitation. The foods we eat and most of the materials we use for paper, buildings, and clothing are dependent on soils. We have to know about soil to properly manage environmental resources. Understanding soil is important in knowing where to build our houses, roads, and buildings and in evaluating their environmental impacts.

Soils are composed of three main ingredients: minerals, organic materials from the remains of dead plants and animals, and pores that may be filled with air or water. A good-quality soil for growing plants should have about 45% minerals, 5% organic matter, 25% air, and 25% water. Soils are dynamic and change over time. Some properties, such as soil moisture content, change very quickly (within hours), while some changes, such as mineral transformations, occur very slowly (over thousands of years).

Soil formation (pedogenesis) and the properties of soils are the result of five key factors. These soil-forming factors are as follows:

1. **Parent material:** The material from which the soil is formed. Soil parent material can be bedrock, organic material, or surface deposits from water, wind, glaciers, or volcanoes.

2. **Climate:** Heat and moisture break down the parent material and affect the speed of soil processes.

3. **Organisms:** All plants and animals living on or in the soil. The dead remains of plants and animals become organic matter in the soil, and the animals living in the soil affect the decomposition of organic materials.

4. **Topography:** Topography impacts soil development. For example, soils on the side of hills tend to be shallow, while soils in valleys tend to be deeper, darker, and contain more horizons.

5. **Time:** All of the aforementioned soil-forming factors assert themselves over time—from hundreds to tens of thousands of years.

SOIL PROFILES
Due to the interaction of the five soil-forming factors, soils differ greatly. Each soil on the landscape has its own unique characteristics. A cross-sectional view of soil is called a **soil profile**, which can be obtained by extracting a core sample of soil or it can be exposed by a soil pit. The soil profile can be used to determine the properties of a soil and the best use of the soil. Every soil profile is made up of layers called **soil horizons**. Horizons can be identified by changes in color or texture compared to adjacent horizons. Horizons are labeled based on their properties (Figure 21.1).

O HORIZON: The O horizon is made up of organic material and found at the soil surface.

A HORIZON: The A horizon is commonly called the topsoil and is the first mineral horizon in the soil profile. It is mostly made up of sand, silt, and clay particles, but also contains some decomposed organic material.

B HORIZON: The B horizon is composed of minerals that undergo chemical and physical weathering. Weathering causes changes in soil color, texture, and structure. The B-horizon is often rich in clays, iron, and aluminum.

C HORIZON: The C horizon is the parent material from which the above horizons have formed.

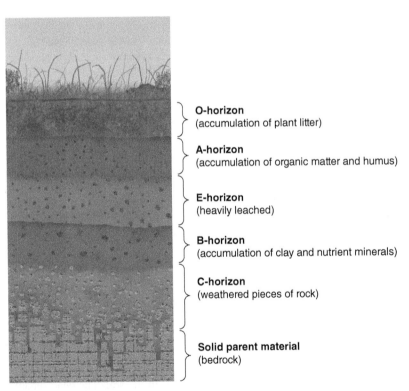

FIGURE 21.1 Soil profile

Each horizon has its own chemical and physical properties. Horizon depths can vary greatly depending on flooding and other factors.

Source: Raven, P.H., L.R. Berg, and D.M. Hassenzahl. 2010. *Environment.* 7th ed. John Wiley & Sons, New York (p. 331).

In soil sampling, every geotechnical engineer and soil scientist knows that augers give highly disturbed samples. However, augers are useful to reach certain depths from which a relatively undisturbed sample can be taken using a split spoon or some type of thin wall tube. Test pits are best for visualizing undisturbed structure and consistency in soil horizons. Typically, test pit log sheets record strata changes and water observations by depths. Soil colors are noted, especially the presence of gray color, which is indicative of the water table. The strata (soil horizons) are also characterized according to texture, structure, consistency, and amount of roots and rocks.

Under some conditions, pits are not dug. Engineers prefer soil borings over pits when the *in situ* (its original place) strength is important, and standard penetration tests can be made at various depths in the boring. This is useful for designing soil-supported foundations.

In this lab, we will excavate a soil test pit and identify soil horizons and determine soil color, soil texture, and the percentage of roots and rocks.

Color: The color of the soil changes depending on how much organic matter is present and the kinds of minerals it contains. Soil color will differ depending on moisture content, and the color can often indicate whether the soil has been saturated with water.

Texture: The texture is the amount of sand, silt, and clay particles in the soil and can be determined by how the soil feels. Sand is the largest size particle in the soil and feels gritty. Silt feels smooth or floury. Clay feels sticky.

MATERIALS

- Nails, 10

- Meter stick or tape

- Munsell color book

- Shovel

- Squirt bottle or water bottle with sprayer

- String and string level

- Sturdy shoes and outdoor clothes

- Tarpaulin

- Trowel

- Water bottle

TASKS

1. Dig a pit 1 m deep and 1 square meter. A pit this big is needed to easily observe all types of soil horizons. Note that you will need to reclaim the site, so be neat and careful with the removed soil and sod, which should be placed on the tarpaulin next to the pit.

2. Starting at the top of the soil profile, observe the profile closely to determine where the different horizons occur. Look carefully for any distinguishing characteristics such as different colors, roots, and the amount and size of rocks.

3. Mark the horizon boundaries with nails, and measure the top and bottom depths for each horizon, in centimeters, and record on a piece of graph paper (depict the horizons, measurements, and observations).

4. Assign the appropriate label (O, A, B, C) to each horizon.

5. Characterize each horizon for (a) color using a Munsell color book, (b) texture using the soil texture flowchart and the **soil textural triangle** (see Figures 21.2 and 21.3), and (c) the percentage and size of roots and rocks.
   A. **For color:** Take a sample from the horizon and moisten it slightly with water from your water bottle. Hold the color chart next to the soil and determine which color matches the color of your soil. Stand with the sun over your right shoulder so that the sun shines on the color chart and your soil sample. Record the color on your soil characterization sheet. Sometimes, a soil horizon may have more than one color. Record all colors.
   B. **For texture:** See Figures 21.2 and 21.3.
   C. **For roots and rocks:** Estimate the percentage and approximate size of roots and rocks.

6. Other site information—Spend a few minutes recording the details of the site.
   A. Record the dominant vegetation at the site.
   B. Record the percent slope of the land (rise ÷ run).
   C. Record the location of the soil pit using major features, such as buildings or utility poles.
   D. Record any other distinguishing properties of the site.
   E. Record time of day, date, approximate temperature, and weather.

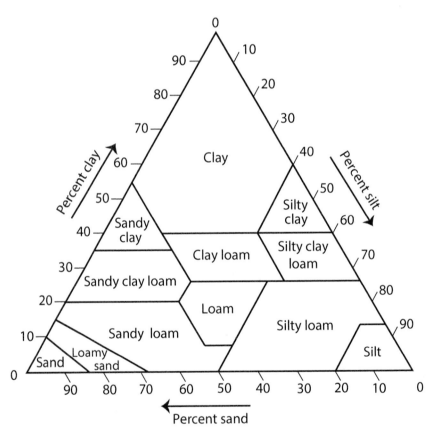

FIGURE 21.2 Soil texture triangle

This triangle illustrates the 12 USDA soil classifications and is commonly used in classifying soil textures.

Source: Soil Survey Division Staff. 1993. Soil Survey Manual. Soil Conservation Service. U.S. Department of Agriculture. Handbook 18. U.S. Government Printing Office, Washington, DC.

   F. Document evidence of previous disturbance (fill, burrowing) in your pit and on the site.
   G. What biotic activity can you see in the walls of your test pit?
   H. Did you find evidence of a current or ancient water table?

8. Completely restore the site so that in a few weeks no one can tell that a pit was dug.

Step 1

A. Obtain and moisten a sample from each Horizon.
B. Place some soil from a horizon (about the size of a small egg) in your hand.
   i. Using a spray bottle, moisten the soil.
   ii. Let the water soak in and then work the soil between your fingers until it is moist throughout.
   iii. Try to form a ball.
   iv. If the soil forms a ball, go to **Step 2**.
   v. If the soil does *not* form a ball, go to **Step 5**.

Step 2

A. If the soil: • Is really sticky • Is hard to squeeze • Stains your hands • Has a shine when rubbed • Forms a long ribbon (5+ cm) without breaking Classify it as *clay* and go to **Step 3**; otherwise, go to **B**.	**C.** If the soil: • Is soft • Is smooth • Is easy to squeeze • Is at most slightly sticky • Forms a short ribbon (< 2 cm) Classify it as *loam* and go to **Step 3**; otherwise, go to **D**.
B. If the soil: • Is somewhat sticky • Is somewhat hard to squeeze • Forms a medium ribbon (between 2 and 5 cm) Classify it as *clay loam* and go to **Step 3**; otherwise, go to **C**.	**D.** If the soil forms a ball but no ribbon, go to **Step 4**.

Step 3

Refine the initial soil texture classification from **Step 2** for relative amounts of *sand* and *silt*. Wet a small pinch of soil in your palm and rub it with a forefinger. If the soil: feels gritty, go to **E**; very smooth with no gritty feeling, go to **F**; or only a little gritty, go to **G**.	**E.** Add the word *sandy* to the initial classification. Soil texture is (select one) • Sandy clay • Sandy clay loam • Sandy loam Soil texture is complete.

continued

F. Add the word *silt* or *silty* to the initial classification.
 	
Soil texture is (select one)
   ● Silty clay
   ● Silty clay loam
   ● Silt loam
Soil texture is complete.
	G. Leave the original classification of (select one)
	   ● Clay
	   ● Clay loam
	   ● Loam
	Soil texture is complete.

Step 4

Step 5

(Test for *loamy sand* or *silt*)
 	
If the soil:
   ● Forms a ball
   ● Forms no ribbon
   ● And is:
 	
H. Very gritty (otherwise go to **I**)
Soil texture is *loamy sand*
Soil texture is complete.
 	
I. Very soft and smooth with no gritty feeling.
Soil texture is *silt*.
Soil texture is complete.
	Test for Sand
	
	If the soil does not form a ball and instead falls apart in your hand,
	
	Soil texture is *sand*.
	Soil texture is complete.

FIGURE 21.3 Steps for classifying soil texture

Working your way through these steps with a soil sample should lead to successful classification of texture.

Source: Soil Survey Division Staff. 1993. Soil Survey Manual. Soil Conservation Service. U.S. Department of Agriculture. Handbook No. 18. U.S. Government Printing Office, Washington, DC.

WRITE-UP

Your lab report will be a technical report. Use the following format:

Title:

   I. Introduction: Background information explaining the importance of soil to the environment.

   II. Methods: Explain what and how you collected your data.

   III. Results: Present your findings; be sure to use a table or graph.

   IV. Discussion: Interpret your results.

   V. References Cited

Climate Change and Sea Level Rise

OBJECTIVES

- Determine the existence of the greenhouse effect.

- Explore the connection between global warming and sea level rise.

- Describe the role of modeling in assessing vulnerability for sea level rise.

- Investigate a current location facing potential negative impacts from sea level rise.

KEY TERMS AND CONCEPTS

✓ Albedo effect

✓ Greenhouse effect

✓ Positive feedback loop

✓ Thermal expansion

TIME REQUIRED: 1.5 hours for experimentation and 0.5 hour for Internet/Library research.

INTRODUCTION

In addition to photosynthesis, the sun plays a critical role in warming Earth sufficient to support life. Life has evolved over billions of years based on the sun's heat input (Figure 22.1), which is natural global warming, or the **greenhouse effect**. The greenhouse effect occurs when atmospheric greenhouse gases, especially carbon dioxide (CO_2), absorb sunlight and solar radiation that have bounced off the Earth's surface. Without greenhouses gases, radiation would escape into space, but instead, some of the heat is trapped to help warm Earth (see Figure 22.2) sufficient for life to exist. If, however, the climate changes too rapidly, there is the potential for many known and unknown negative impacts on life.

Human activities involving CO_2 emissions are increasing the greenhouse effect and thus impacting global climate systems. The concentration of CO_2 in the atmosphere has increased steadily since the rise of our reliance on fossil fuels, beginning with the Industrial Revolution, coupled with massive land-use changes such as deforestation. As a result (see Figure 22.3), the rise of the atmospheric concentration of CO_2 over the past 70 years has been dramatic. The National Oceanic and Atmospheric Administration (NOAA) estimates that the rate of increase is 200 times faster than the last sustained period of increase between 17,000 and 11,000 years ago (NOAA, 2016), In 2013, CO_2 levels surpassed 400 parts per million (ppm) for the first time in recorded history (NASA, 2016), Although 400 ppm (or 0.04%) seems a very small concentration, every year it is increasing by more than 2 ppm (NASA, 2016).

Too much CO_2 can trap too much heat in the Earth's atmosphere. An analogy is using a thick wool blanket on a hot summer night. Increased atmospheric heat has many implications. Increased heat melts glaciers and ice sheets more than normal. Ice and snow provide an **albedo effect** (the amount of solar energy reflected from Earth back into space). However, as land-based ice and snow melt, their contribution to albedo shrinks. A reduction in reflective surfaces means greater absorption of heat and thus an increase in atmospheric warming. This is an example of a positive feedback loop (**positive feedbacks** amplify the effect of the input, which may not actually be beneficial).

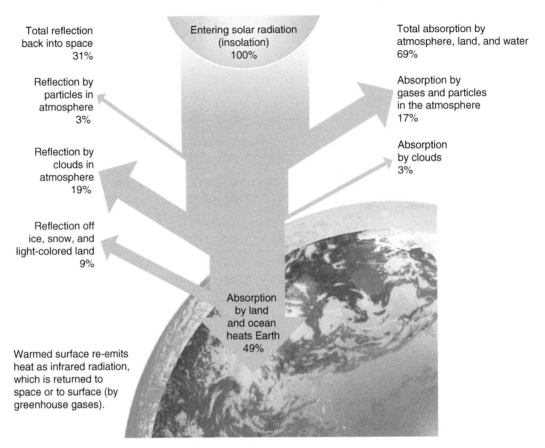

FIGURE 22.1 Fate of solar radiation that reaches Earth

Most of the energy the sun produces never reaches Earth. The solar energy that does reach Earth warms the planet's surface, drives the hydrological cycle and other biogeochemical cycles, produces our climate, and powers almost all life through the process of photosynthesis.

Source: Berg, L.R., D. M. Hassenzahl, M.C. Hager. 2013. Visualizing Environmental Science, 4th ed., Wiley.

The increased atmospheric temperatures and melting ice will increase the sea level because there is more water entering the oceans and because of thermal expansion. **Thermal expansion** of oceans occurs because warmer water expands. Thus, as the atmospheric temperature rises, and oceans absorb heat from the atmosphere, oceans warm and thus expand. According to the US EPA (2017), averaged over all of the world's oceans, the absolute sea level has risen 0.06 in (1.52 mm) per year from 1880 to 2013. However, the rate of sea level rise since 1993 is increasing nearly twice as fast as the long-term trend and is 0.11 to 0.14 in (2.79 to 3.57 mm) per year (NASA, 2016).

Oceans themselves have much variation in their composition, movements, and other characteristics. These too are affected by climate change. Temperature and salinity changes will have compound impacts. For example, high salinity and low temperature produce dense seawater that sinks to the ocean bottom, where it flows in its own layer of current, with dynamic consequences.

In this lab, you will first take measurements demonstrating the greenhouse effect. Then, you will test for thermal expansion to determine if the volume of water increases with temperature. Finally, you will explore models mapping vulnerability from sea level rise.

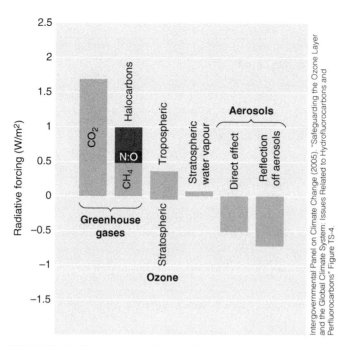

FIGURE 22.2 Enhanced greenhouse effect

The buildup of carbon dioxide (CO_2) and other greenhouse gases warms the atmosphere by absorbing some of the outgoing infrared (heat) radiation. Some of the heat in the warmed atmosphere is transferred back to Earth's surface, warming the land and ocean.

Source: Berg, L.R., D.M. Hassenzahl, M.C. Hager. 2013. Visualizing Environmental Science, 4th ed. Wiley.

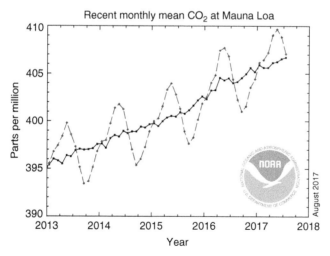

FIGURE 22.3 Monthly mean carbon dioxide measured at Mauna Loa Observatory, Hawaii (NOAA, 2016)

The dashed line represents the monthly mean values, centered on the middle of each month. The solid line represents the same, after correction for the average seasonal cycle.

MATERIALS

- Florence flask, 250 mL
- Food coloring, blue, green, or red
- Hotplate
- Marker
- Ruler (to mark mm on expansion tube)
- Safety goggles
- Salt, sodium chloride, approximately 50 g
- Scale, in grams
- Flask stopper made of rubber, with one hole (two holes if also inserting a thermometer)
- Thermometer, infrared
- Tongs, to move the flask
- Tube, one thin glass or plastic tube, at least 25 cm long
- Water, approximately 1,000 mL

TASKS

1. THE GREENHOUSE EFFECT

Low clouds and humid air have the strongest greenhouse effect, whereas a very dry, clear atmosphere has the weakest greenhouse effect.

A. Develop an alternate *and* null hypothesis predicting the temperature difference between (a) a clear sky and (b) a cloud.
H_O:
H_A:

B. Using an IR thermometer, take three measurements of a clear (no clouds) portion of the sky. Record these temperatures in Table 22.1 and then calculate the mean.

C. In the same sky, using the IR thermometer, take three measurements of a cloud. Record these temperatures in Table 22.1 and calculate the mean.

D. What is the percent difference between the two mean temperatures?

E. Discuss the results in relation to your hypothesis. What is your finding on the existence of the greenhouse effect?

CALCULATING PERCENT DIFFERENCE

$$\frac{\text{1st Value} - \text{2nd Value}}{(\text{1st Value} + \text{2nd Value})/2} \times 100$$

Table 22.1   Data Collection Table for Greenhouse Effect

Temperature	Clear sky	Cloud
#1		
#2		
#3		
Mean		

2. THERMAL EXPANSION OF WATER

A. Develop an alternate *and* null hypothesis predicting what will happen to the volume of salt water when heated (i.e., will the volume expand, stay the same, or decrease?).

H_O1:

H_A1:

B. What is the natural, average salinity level of the ocean nearest to you (be sure to cite your source)? It is probably expressed as a percent or parts per thousand (ppt). Using table salt (sodium chloride), create artificial seawater (it is artificial because the oceans contain minerals, organic matter, pollutants, and other materials that will not be in your solution) using the formula in the sidebar. Then, add food coloring so that it is a vibrant color.

C. You will be constructing an apparatus similar to the one depicted in Figure 22.4. First, put on your safety goggles (leave them on during the entire Task 2 portion of this lab) then fill the flask with your colored artificial saltwater. Second, carefully insert the glass or plastic tube into an open hole of the rubber stopper. Make sure there is a tight seal between the tube and the hole in the stopper. Next, using the ruler and marker, starting at the top of the stopper, carefully and accurately mark centimeters (cm) on the entire length of the glass/plastic tube. Third, if you are using a glass thermometer, carefully insert it into the other hole in the stopper again making sure there is a tight seal. (If you are using the IR thermometer, skip this step.) Fourth, insert the rubber stopper with the tube (and thermometer) into the top of the flask. The stopper has to have a tight seal so that when the saltwater expands, it does so only through the tube. Finally, make sure the level of the saltwater is even with the top of the stopper. If there is too little, add some more, but make sure the level is even with the top of the stopper.

CREATING ARTIFICIAL SALTWATER

$$\frac{xx \text{ gms of salt}}{1,000 \text{ g } H_2O} \times 100\% = X.X\%$$

Seawater salinity should range between 34 and 37 ppt. $(34/1,000) \times 100 = 3.4\%$

Thus, for every 1,000 mL of water, which is the same as milligrams, add 34 to 37 mg of salt.

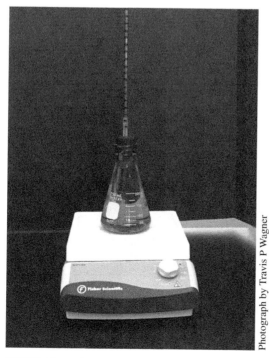

Photograph by Travis P Wagner

FIGURE 22.4   Apparatus for testing for thermal expansion

D. Place the flask on the hotplate. Record the temperature into Table 22.2. Turn on the hotplate to a setting of 125°C. Record the temperature and water level every 2 minutes into Table 22.2. Continue doing this until the temperature has increased by a total of 40°C over the starting temperature.

E. Based on your data in Table 22.2, create a line graph that depicts the correlation with temperature and volume expansion of the saltwater.

Table 22.2 Data Collection Table for Thermal Expansion of Salt Water

Time (mins)	Temperature (°C)	Water level (cm)
0		
2		
4		
6		
8		
10		
12		
14		
16		
18		
20		

F. Explain the correlation between temperature and the volume of salt water.

3. MODELING THE VULNERABILITIES TO AND IMPACTS FROM SEA LEVEL RISE

A. Go to the U.S. National Oceanic and Atmospheric Administration's (NOAA) Sea Level Rise Viewer: https://coast.noaa.gov/slr/beta/#/layer/slr. If you live or attend college at a coastal city, input the city. If not, your instructor will suggest a variety of cities to choose from. Make sure the SEA LEVEL RISE button has been selected. In the column to the left labeled water level, the MHHW (Mean Higher High Water) should be at zero, select 1 ft and record what you see. Next, slowly move the slide all the way up to 6 ft pausing at each ft mark and record what you see. Then move the sea level slide back to MHHW.

B. Now, make sure the VULNERABILITY button has been selected. In the column to the left labeled water level, the MHHW should be at zero, select 1 ft and record what you see. Next, slowly move the slide all the way up to 6 ft pausing at each ft mark and record what you see.

C. Using the NOAA Sea Level Rise map, or your instructor's guidance, select a city that is currently facing the potential impacts of sea level rise (e.g., Boston, MA; Miami Beach, FL; New Orleans, LA; Venice, Italy; etc.). What is the population of the city? Based on the map, what are the projected impacts relating to various scenarios of sea level rises.

D. Go to the city's website and try to locate a published plan regarding the impacts of climate change. If there is a plan, to what degree is sea level rise perceived to be a potential problem for the city. Is the problem minor, moderate, major, or serious? Be sure to provide a complete citation for the city's report.

This is a formal lab write-up. Be sure to use the following headings:

Title	A short, concise title that indicates the subject matter.
I. Introduction	A brief paragraph as to what the lab is about. A brief discussion of the relevancy of this lab activity. For example, discuss the environmental implications of the sea level rise and the direct and indirect roles humans play in causing sea level rise. In this section, it is imperative to cite references. Finally, this section ends with each of your stated hypotheses.
II. Methods	Describe how you collected the data for each experiment.
III. Results	What were your results? Insert the table of the results for Task #1 and the line graph as discussed above.
IV. Discussion	State and explain if you rejected or accepted each hypothesis and explain why. Then discuss the following:

- Discuss how your selected city from Task #3 likely will be impacted based on your results of the thermal expansion experiment.
- How vulnerable is your selected community to various amount of increase in sea level rise?
- Discuss whether it is better to reduce the *causes* of sea level rise or to reduce the *consequences* (impacts) of sea level rise?

V. References Cited An alphabetical list of all cited references in proper format.

NASA. 2016. Global Climate Change: Vital Signs of the Planet. Available at https://climate.nasa.gov/climate_resources/24 (verified 17 August 2017).

NOAA (National Oceanic and Atmospheric Administration). 2016. Record Annual Increase of Carbon Dioxide Observed at Mauna Loa for 2015. Climate Research News and Features, US Department of Commerce. Available at http://www.noaa.gov/news/record-annual-increase-of-carbon-dioxide-observed-at-mauna-loa-for-2015 (verified 17 August 2017).

US EPA. 2017. Climate Change Indicators: Sea Level. Available at https://www.epa.gov/climate-indicators/climate-change-indicators-sea-level (verified 17 August 2017).

Reducing the Generation of Solid Waste

OBJECTIVES

- Explain and apply the waste management hierarchy.

- Collect, categorize, and analyze recycling materials.

- Assess the campus community's knowledge of recycling.

KEY CONCEPTS AND TERMS

✓ Compost

✓ Municipal solid waste

✓ Pollution prevention

✓ Pollution Prevention Act

✓ Recycle

✓ Single stream

✓ Source reduction

✓ Reuse

✓ Waste Management Hierarchy

TIME REQUIRED: 1.5 to 2 hours.

INTRODUCTION   Americans are the most wasteful society on the planet. In 1960, the U.S. per capita waste generation rate was 2.7 lb/person/day. Currently, we generate approximately 4.4 lb/person/day of **municipal solid waste** (MSW), which is everyday trash, garbage, rubbish, and refuse from households and institutions and includes paper, yard waste, plastics, metals, wood, glass, textiles, food waste (US EPA, 2016). That means each year the average person generates 1,160 lb of MSW. When we dispose of MSW, we are throwing away items that still have value with regard to resources and/or energy potential. An approach to reduce the amount of waste generated and/or disposed is to adopt and apply the waste management hierarchy.

As shown in Figure 23.1, the **Waste Management Hierarchy** is a management policy/strategy based on the concept of **pollution prevention**, which was established by the U.S. Congress in Section 6602(b) of the **Pollution Prevention Act of 1990**. In this Act, Congress established a national policy that focused on reducing pollution before it is created as opposed to after it is created (a.k.a. end-of-pipe management), which has been the dominant approach. To specifically address MSW, the waste management hierarchy was created (see Table 23.1), which focuses on a hierarchical management strategy designed to reduce the generation of MSW and also the reliance on disposal by employing the strategy at each step in the hierarchy.

As a society, we seem to have difficulty in the simple act of recycling (let alone reducing and reusing waste). It is common practice to combine all our solid waste and let someone else worry about sorting it for processing or reclamation. Yet it generally involves more money, energy, and resources to separate trash after it is generated compared to properly sort it before it is recycled.

Much of our solid waste is "disposed" of in landfills or is incinerated. We bury or burn many things that could be recycled. For example, many states still landfill electronic waste, yet

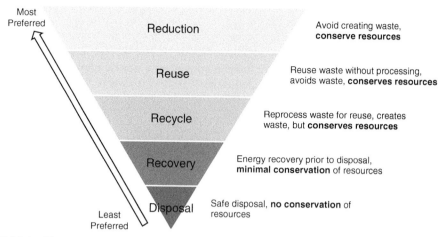

FIGURE 23.1   The waste management hierarchy

As depicted, the waste management hierarchy is an inverted pyramid that provides a hierarchical approach to manage MSW with the most preferred approach at the top and the least preferred approach at the bottom. The point of using the inverted pyramid is to demonstrate conceptually the amount of MSW that could be recovered at each step and that if all preferred options are taken, a comparably small amount would need to be landfilled.

Table 23.1   Strategy Step, Description, and Examples of the Waste Management Hierarchy

Strategy	Description	Examples
Source reduction	Waste should be *prevented* or *reduced* at the source whenever feasible.	Renting/leasing, using concentrated detergents, purchasing materials in bulk, not buying products, purchasing more durable products.
Reuse/recycling	Waste that cannot be prevented or reduced should be *reused/recycled* in an environmentally safe manner whenever feasible.	Donating items; reuse grocery bags; repair items; recycle aluminum, paper, and cardboard, which are turned into new products.
Composting	The organic components of waste should be composted for beneficial reuse.	Turning food waste into a nutrient source for your garden.
Resource recovery	For remaining waste, the energy should be extracted from the waste through combustion to generate electricity or heat.	Burning nonrecyclable waste to capture the heat value to create electricity.
Landfilling with methane recovery	The least preferable approach is to landfill waste, but when this occurs, methane should be captured.	Captured methane gas generated by decomposition of the organic portion of waste can be used to power vehicles or generate electricity.
Landfilling without methane recovery	The least preferable management method is one that fails to capture methane from landfill waste.	The old way of managing waste.

"e-waste" contains ordinary metals such as steel and aluminum; precious metals such as gold and platinum; and toxic heavy metals such as mercury, lead, cadmium, and nickel. All these materials could be valuable raw materials reducing our need to mine for metal ores.

In recent years, there has been a sustained effort to divert waste away from disposal and recapture resources or energy through recycling. The United States currently recycles about 35% of our MSW, which represents a doubling since 1990. Recycling is less preferable than **source reduction** or reuse, but much better than simply dumping, hiding, burning, or burying the waste.

1. CAMPUS RECYCLING

Many campuses employ some level of recycling, which can be segregated on-site (paper in dedicated bin, metal in a dedicated bin, etc.) or collected as **single stream** (e.g., single sort or no sort) where all recyclable materials go into *the same bin* together. From campus, the collected recyclables are generally taken to a nearby materials recovery facility where they are further sorted based on the market demands for various commodities. In general, the following materials are collected for recycling on campuses:

- **Paper, paperboard, cardboard:** Office papers, newspapers, paper envelopes, books and magazines, cereal and frozen meal boxes, clean paper plates, paper cups, drink cartons, and cardboard and pizza boxes.

- **Plastic:** Any *empty rigid* container marked with a recycling triangle (resin identification codes #1 to 7) such as frozen meal trays, cups, to-go containers, and detergent and shampoo bottles. Increasingly, plastic grocery bags (#4 and #6) are not collected because they are often not recycled.

- **Metal:** Empty aluminum, tin and steel food cans, and aluminum foil.

- **Glass:** All colors of empty glass bottles and jars, lids removed.

Depending on the campus, the following materials often are not collected for recycling due to a variety of reasons:

- Napkins, paper towels, and tissues

- Expanded polystyrene (Styrofoam™) including cups, plates, clamshell containers, and packing peanuts, and packaging

- Plastic wrap

- Granola and candy bar wrappers

- Plastic grocery bags, trash bags, and potato chip bags

- Wax paper

- Disposable gloves, textiles

- Food scraps (these can be composted)

In this lab, you will be characterizing collected recycling materials to assess how well campus community members are doing with regard to following the waste management hierarchy. You also will be conducting a visual inspection of collection containers and conducting a short survey.

MATERIALS	

- Gloves

- Plastic buckets, 5-gal, enough to temporarily store categories of recyclables

- Recycling (recyclables from a collection container)

- Safety goggles

- Scales, to weigh ounces or grams

- Survey sheets

- Tarp

- Tongs (ones used for grilling work well)

1. RECYCLING CHARACTERIZATION

The first task is to conduct a *characterization of recyclables*, a process in which the composition of the recycling containers is sorted, categorized, and analyzed to assess the degree to which the campus community is following the waste management hierarchy. The following can be done by teams or as a class. Any person working with the recyclables should wear gloves and safety goggles:

A. Determine the major categories of materials collected for recycling on your campus. Based on the answer, set up and label 5-gal buckets, one for each major category. (Be sure to also have a bucket for trash, or nonrecycled materials.)

B. Empty the contents of the bag of recyclables onto the tarp. As a group, using the tongs, carefully sort the recyclables into the categories and place the materials into their respective buckets. Once you've completed the sorting process, weigh each category *and* estimate the volume within the bucket to calculate the *percentage* of each category, and then record these in Table 23.2. (Be sure to weigh only the contents and not the bucket itself.)

C. Create a figure with two pie charts to depict the proportion of the major categories of recyclables; one pie chart depicting recyclables by weight and the other one depicting recyclables by volume.

Table 23.2 Recyclables Characterization Data Sheet

Material	Weight (oz.)	Percentage of total (%)	Volume (gal)	Percentage of total (%)
Paper Cardboard, mixed paper, newspapers, paper coffee cups, pizza boxes				
Returnable containers (bottle bill states only) Bottles and cans subject to a deposit/refund				
Plastic Plastic containers, tubs, trays, coffee lids				
Glass Bottles and jars				
Metal Aluminum foil, steel cans, and other steel products				
Other Recyclable materials not covered in the above categories				
Waste Nonrecyclable items				
Total				

2. CONTAINER ASSESSMENT

In your team, conduct a thorough walkthrough of a selected building by walking down each hall on each floor to answer the following:

A. Count the number of waste, recycling, food waste/organics (if applicable), and beverage collection (if applicable) containers throughout the entire building and input into Table 23.3.

B. You characterized a small amount of recycling materials collected (e.g., recyclables from one container, from one building, from one day). Based on what you collected and the total number of recycling containers, estimate the amount of recycling materials produced in your building for the entire school year.

C. Visually inspect the containers (don't root through or "dive" into containers and risk personal injury). Input all the data into Table 23.3. Be sure to take photos to document what you found.
  - **I.** Calculate the frequency (percent) of "contamination" (inappropriate materials for the collection container such as recyclables in the trash) in percent for *each* collection container type.
  - **II.** Identify and describe the specific types of inappropriate materials in the waste container.
  - **III.** Identify and describe the specific types of inappropriate materials in the recycling container.
  - **IV.** Provide photos of examples of contamination.

D. During your inspection of the building:
  - **I.** Count the number and type of posters, signs, and education materials that explain why we should recycle, where we should recycle, and how (e.g., the proper bin for the various materials).
  - **II.** Provide photos of some sample signs.
  - **III.** Comment on what could be done to improve the messaging of the signs.

E. Go to a food service operation (e.g., snack bar, cafeteria) and conduct a visual inspection.
  - **I.** Is the operation attempting to promote source reduction, reuse, or recycling (If yes, how; if no, why)?
  - **II.** What percentage, roughly, of the food containers/packaging are recyclable on campus?

Table 23.3 Container Assessment Data Collection Sheet

Location	Container bin type	*Contaminated?	Inappropriate materials
Near room 1xx Smith Hall	Trash	Yes	Pizza boxes, paper
	Recycling	Yes	Plastic bags, food waste
	Food waste	No	Food wrappers
	Beverage containers	Yes	Plastic cups

*Contamination rate refers to the percentage of recyclables found in the waste container and the percentage of waste found in the recycling container.
The shaded row is an example of a hypothetical response.

Table 23.4   Results of the Campus Community Knowledge Assessment Survey (N = x)

Item	Yes	No	Not sure	Is it recyclable on campus?*
Pizza box				
Milk carton				
Styrofoam™				
Napkins				
Potato chip bag				
Coffee cup from campus kiosk				
Coffee cup lid from campus kiosk				
Shampoo bottle				
Candy wrapper				
Plastic grocery bag				

*Is it recyclable at your campus? This will help you assess the accuracy of the responses.

3. CAMPUS COMMUNITY KNOWLEDGE ASSESSMENT[1]

A. Select a random sample (a minimum of 10) of students, staff, or faculty using the Campus Recycling Awareness Survey presented at the end. Compile the results into Table 23.4 and be sure to include the "N" (the number of respondents) in the title.

B. Based on your survey results, identify where the respondents were incorrect (e.g., did they say something was not recyclable but in fact it is?) *and* what overall conclusions can you draw regarding students' environmental awareness?

Your lab report will be a technical report. Use the following format:

WRITE-UP

Title:

I. **Introduction:**   Background information explaining the role and importance of recycling. Describe the recycling program on your campus including the major categories of materials recycled.

II. **Methods:** Explain who, what, and how you collected your data through your waste characterization, visual assessment, and surveying.

III. **Results:**   Present your findings; be sure to use tables (e.g., container assessment results, survey results) and figures (e.g., pie charts).

IV. **Discussion:**   Interpret your results by assessing actual practices and expressed knowledge. Be sure to discuss your results in relation to the waste management hierarchy. Also, make three specific recommendations to improve the management of waste by moving up the hierarchy based on practices adopted by other campuses.

V. **References Cited**

---

[1] Before engaging in any survey, contact your campus Institutional Review Board to determine any applicable requirements and/or approvals.

REFERENCE

U.S. Environmental Protection Agency (US EPA). 2016. Advancing Sustainable Materials Management: Facts and Figures Report [Online]. Available at https:// www.epa.gov/smm/advancing-sustainable-materials-management-facts-and-figures-report (verified 14 August 2017).

1. CAMPUS RECYCLING AWARENESS SURVEY

Hello. We are conducting an anonymous survey for a school project. Completion of the survey will take approximately 4 minutes. We will keep your answers confidential and you will not be identified as a respondent.

Question			
How often to do you recycle on campus?			
What percent of all waste generated at the university is recyclable?			
What waste materials generated at the university do you believe are recyclable?			
Do you believe there are appropriate and sufficient number of containers for recycling?			
Item	**Yes**	**No**	**Not sure**
Pizza box			
Milk carton			
Styrofoam			
Napkins			
Potato chip bag			
Coffee cup from campus kiosk			
Coffee cup lid from campus kiosk			
Shampoo bottle			
Candy wrapper			
We have found numerous instances of trash in recyclable containers and recyclables in trash containers. What two suggestions can you make to reduce this practice?			

Please complete the information below as it applies to you:

Gender: _____Male _____Female _____ Other

Status: _____Part-time student_____Full-time student_____Staff_____Professor _____Visitor

Primary source of news: Television _____ Facebook _____ Twitter _____

Newspaper _____ Other _____

Thank you for your participation

_____.

Reducing Campus Food Waste

OBJECTIVES

- Explain and apply the Food Recovery Hierarchy.

- Collect, categorize, and analyze recycling materials.

- Assess the campus community's knowledge of food waste.

KEY CONCEPTS AND TERMS

✓ Composting

✓ Food insecurity

✓ Food Recovery Hierarchy

✓ Food waste

✓ Source reduction

TIME REQUIRED: 1.5 to 2 hours.

INTRODUCTION

According to the US EPA (2017), in 2014, we threw away more than 38 million tons of wasted, excess, and surplus food, at the same time approximately 13% of American households faced **food insecurity** (limited or uncertain access to adequate food). **Food waste** is generated by a variety of sources including food producers, farms, grocers, restaurants, households, and institutions including hospitals, prisons, cafeterias, and college campuses. Food waste includes food that has spoiled, uneaten portions or components of plated food, byproducts of food and beverage processing industries, and foods used to cook other foods (fats, oils, and greases). Food waste constitutes about 21.6% of the amount by weight of waste that is disposed of nationally (US EPA, 2017).

There are increasing actions by the federal, state, and, especially, local governments to reduce the amount of food waste generated and to increase the recovery of food waste. **The Food Recovery Hierarchy** is a management approach based on the concept of pollution prevention (Figure 24.1 and Table 24.1). It focuses on a hierarchical management strategy designed to reduce the generation of food waste and maximize the recovery of the highest value of food at each step in the hierarchy.

The production, distribution, and preparation of food have significant environmental impacts. To produce food, water, fertilizers, feed, pesticides, and fuels are required. And, food needs to be stored (e.g., refrigeration, freezing) prior to use, sometimes for extended periods of time and during transport, all of which required energy. Thus, any food that has been wasted or not been consumed expends energy and resources that could have been put to better use, a foundation of sustainable practices.

The purpose of this lab is to expand your knowledge of the campus food waste reduction and management program while examining how well the program works and the awareness of campus community members regarding food waste on campus.

FIGURE 24.1   The Food Recovery Hierarchy

The U.S. EPS Food Recovery Hierarchy provides a hierarchical approach to manage food waste with the most preferred approach at the top and the least preferred approach at the bottom.

Table 24.1   Strategy Step, Description, and Examples of the Food Waste Recovery Hierarchy

Strategy	Description	Examples
Source reduction	Preventing the generation of food waste	Not overordering, not using trays in cafeterias, proper storage, appropriate serving sizes, reuse scraps for other food items (e.g., soups)
Feed hungry people	Donating or selling food to people in need	Food pantries, homeless shelters, charities, and individuals
Feed animals	Donating or selling food to feed animals	Livestock farms, zoos, refuges, and pets
Industrial uses	Producing of biofuels and bio products for industrial uses	Convert food into biodiesel, methane, cosmetics, soap, etc.
Composting	Aerobic treatment of food waste	Composting food wastes creates a product that can help improve soils
Disposal	The least preferable option	Landfilling or burning for energy recovery

MATERIALS

- Gloves
- Plastic buckets, 5-gal, enough for each major category of food waste and one for nonfood waste
- Food waste (food waste from a collection container)
- Safety goggles
- Scales, to weigh ounces or grams
- Tarp
- Tongs, such as those used for grilling

1. FOOD SERVICE ASSESSMENT

A. Go to the cafeteria or other food service location and select 10 food items (i.e., individual items such as a yogurt or a banana). For each item that is containerized or bagged, what is its shelf life (in weeks or months) based on the buy-by, sell-by, or consume-by dates? For each perishable item (e.g., fruits, vegetables, or meats), what is its likely shelf life (in days) based on guidelines or appearance of the food item? What is its country of origin? If in the United States, what is the state or territory of origin? Approximately how many miles is the country from your campus? Input this data into Table 24.2.

B. What percentage of food served at the cafeteria is defined as local (e.g., produced within 100 mi of campus)?

C. What percentage of food served at the cafeteria is defined as organic?

D. What percentage of food served at the cafeteria is grown on campus?

E. Identify which common source reduction practices are used at the cafeteria:
   i. Are trays used?
   ii. Are food portions controlled?
   iii. Are food preparation scraps reused (e.g., soups, casseroles)?

F. Identify which common strategies are employed to reduce the disposal of food waste at the cafeteria:

   i. What percentage of food is donated?
   ii. What percentage of food is sent to a farm for animal feed?
   iii. What percentage of food is composted? If yes, how much of the food is composted on the campus?

Table 24.2 Food Service Assessment Data Sheet

	Item	Shelf life	Country of origin	Miles from campus
1				
2				
3				
4				
5				
6				
7				
8				
9				
10				

2. FOOD WASTE CHARACTERIZATION

The first step is to conduct a *characterization of food waste*, which is a process where the composition of collected food waste is sorted, categorized, and analyzed to identify what and how much food is being wasted and the probable reasons. The following can be done by teams or as a class. Any person involved in the characterization should wear gloves and safety goggles:

A. Using Table 24.3, set up and label 5-gal buckets with the appropriate categories.

Table 24.3 Food Waste Characterization Data Sheet

Food item	Weight (oz.)	Percentage of total (%)
Vegetables		
Uneaten, whole vegetables		
Partially eaten vegetables		
Spoiled vegetables		
Fruit		
Uneaten, whole fruit		
Partially eaten fruit		
Spoiled fruit		
Grains		
Uneaten, whole bread pieces		
Partially eaten bread		
Moldy bread, pasta, pizza, noodles		
Meat		
Uneaten, whole meat		
Partially eaten meat		
Spoiled meat		
Dairy		
Desserts		
Coffee grounds		
Liquids		
Liquids remaining in containers		
Food packaging		
Plastic		
Paper		
Styrofoam™		
Metal		
Other materials		
Utensils		
Food packaging		
Napkins		
Trash		
Other		
Total		-

B. Empty the contents of the provided container of food waste onto the tarp. As a group, using the tongs, carefully sort the food waste into the categories by placing the waste into their respective buckets.

C. Once you've completed the sorting process, weigh each category and calculate the percentage of each category, then record in Table 24.3. (Be sure to weigh only the contents and not the bucket itself.)

    I. Display the data in a figure (i.e., pie chart).

    II. Summarize your findings based on what you found regarding the generation of food waste on campus.

3. CAMPUS COMMUNITY KNOWLEDGE ASSESSMENT[1]

Select a random sample (a minimum of 10) of students, staff, or faculty using the Campus Food Waste Awareness Survey.

Campus Food Waste Awareness Survey

Hello. We are conducting an anonymous survey for a school project. Completion of the survey will take approximately 3 minutes. We will keep your answers confidential and you will not be identified as a respondent.

Question	
What fruit is bought in largest volumes in your dorm room or home?	
What vegetables are bought in largest volumes in your dorm room or home?	
What type of store do you conduct most of your shopping?	
What is the most wasted fresh fruit in your dorm/home?	
What is the most wasted vegetable in your dorm/home?	
What other food products are wasted in your home?	
What is your most common reason for throwing out food?	
How often do you consume food that is past its "best-by" date?	
How often do you eat leftovers that you create?	
What are the three actions that can generally reduce the amount of food waste?	

Please complete the information below as it applies to you:
Gender: _____Male _____Female _____ Other
Status: _____Part-time student_____Full-time student_____Staff_____Professor _____Visitor
Primary source of news: Television _____ Facebook _____ Twitter _____
Newspaper _____   Other _____
Thank you for your participation.

Your lab report will be a technical report. Use the following format:

                                                                               WRITE-UP

Title:

   I. Introduction: Background information explaining the social and environmental importance of food waste. Describe the food waste reduction program on your campus.

---

[1] Before engaging in any survey, contact your campus Institutional Review Board to determine any applicable requirements and/or approvals.

II. **Methods:** Explain who, what, and how you collected your data through your food waste characterization, cafeteria assessment, and surveying.

III. **Results:**   Present your findings; be sure to use tables and graphs.

IV. **Discussion:**   Interpret your results by assessing actual practices and expressed knowledge. Be sure to discuss your results in relation to the Food Recovery Hierarchy. Include three specific recommendations to reduce the generation and/or disposal of food waste by moving up the hierarchy based on practices adopted by other campuses.

REFERENCE

US EPA. 2017. Sustainable Management of Food Basics. U.S. Environmental Protection Agency. Available at https://www.epa.gov/sustainable-management-food.

Compost Facility Planning and Siting

OBJECTIVES

- Interpret standard geographical, historical, and natural resource maps.

- Use a variety of maps to determine what areas are best suited for siting a new compost facility.

- Identify other key factors in siting a new compost facility.

KEY CONCEPTS AND TERMS

✓ Environmental impact assessment

✓ Geographic information system (GIS)

✓ Leachate

✓ Soil survey

TIME REQUIRED: 2 hours.

INTRODUCTION

Environmental science plays a role in planning and developing new public utilities and related infrastructures in landscapes through the description, prediction, and assessment of the range and types of environmental impacts that might occur. Planners use this information to help make decisions. When constructing a new project (e.g., building, road, facility), a fundamental task is to determine the best place to put it. Environmental professionals ask, what is the best location to have an affordable and useful project without too much compromise of environmental quality? The project needs to be near its users and near the resources that serve it while minimizing impacts to human health and the environment.

Environmental professionals who do planning or conduct an **environmental impact assessment** (a process of determining the major physical, biological, ecological, social, economic, and health impacts of a project or policy) begin a project by examining desktop references such as maps. These desktop references are a valuable source of information to catalog natural resources and to assess potential environmental and other impacts from a proposed project. This lab will familiarize you with important maps and provide an opportunity to discover what types of information can be gleaned from them by assessing the environmental viability of a project by using desktop references. Much of this work is done in the environmental consulting world with **geographic information systems (GIS)**, but the principle is the same as with maps.

MATERIALS

- Aerial photographs

- National Wetlands Inventory maps

- Other maps as available (e.g., road map, bedrock geology)

- Ruler, graph paper

- Sanborn Fire Insurance Maps

- Soil survey maps
- Surficial geology maps
- Topographic maps

SOIL SURVEY MAPS	The U.S. Department of Agriculture's National Cooperative Soil Survey (NCSS) is a county-by-county scientific inventory of U.S. soils on nearly all public and private lands. A soil survey includes soil maps and descriptions of each type of soil in the county. Maps show the location of soils in a county. Descriptions of each soil type include the following:

- Depth of each major soil layer
- The ability of water to infiltrate the soil and how easily roots can penetrate
- The rate at which water moves downward through the soil
- How much water the soil can store for plants
- How acid or alkaline the soil is
- The soil's susceptibility to erosion by water and wind

The **soil survey** identifies the suitability of soils for the construction of buildings, roads, septic tank absorption fields, sewage lagoons, compost facilities, ponds, and dikes and levees. Soil information and other historical documents of the NCSS can be accessed online at the USDA NRCS portal: https://www.nrcs.usda.gov/wps/portal/nrcs/main/soils/survey/partnership/ncss

SURFICIAL GEOLOGY MAPS	Surficial geology maps depict surficial rocks and sediments. Because, in most environments, vegetation, soils, and human structures cover the surface, the underlying rocks and sediments are not directly visible or exposed. Surficial geology maps depict the materials approximately 5 ft below the surface. The portal for USGS National Geologic Map data base is at https://ngmdb.usgs.gov/ngmdb/ngmdb_home.html

NATIONAL WETLANDS INVENTORY MAPS	The National Wetlands Inventory (NWI) of the U.S. Fish & Wildlife Service produces information on the characteristics, extent, and status of the nation's wetlands and deep water habitats. Approximately 90% of the lower 48 states and 34% of Alaska have been mapped. NWI maps are used in myriad applications, including planning for watersheds and drinking water supplies, siting of transportation corridors, construction of solid waste facilities, siting of buildings, wildlife habitat identification, floodplain planning, and endangered species recovery. The U.S. Fish & Wildlife Service National Wetlands Inventory portal is at https://www.fws.gov/wetlands

TOPOGRAPHIC MAPS	The U.S. Geological Survey produces topographic maps of various scales (1:24,000 is common). Note that since the scale is a ratio, it does not matter if it is in metric or English units although information is generally given in both. The most distinguishing feature of topographic maps is the use of contour lines to portray the shape and elevation of the land. Topographic maps render the three-dimensional changes of terrain onto a two-dimensional surface. (However, as shown in Figure 25.1, the slope can be determined from topographic maps.) Topographic maps show and

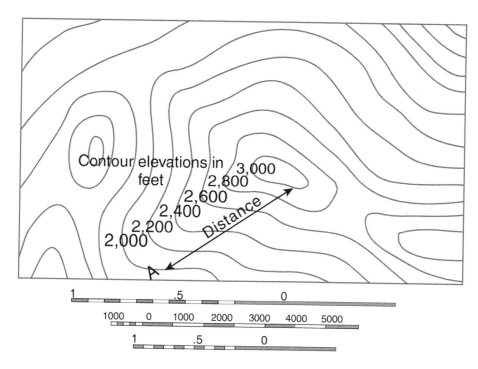

Slope $= \dfrac{\text{Rise}}{\text{Run}}$

% slope $= \dfrac{\text{Rise}}{\text{Run}} \times 100$ **Slope angle** $= \arctan\left(\dfrac{\text{Rise}}{\text{Run}}\right)$

Run = Horizontal distance from A to B = 1 mi = 5,280 ft

Rise = Elevation change = 3,000 − 2,000 = 1,000 ft

% slope $= \dfrac{1{,}000 \text{ ft}}{5{,}280 \text{ ft}} \times 100 = 18.9\%$

Slope angle $= \arctan\left(\dfrac{1{,}000 \text{ ft}}{5{,}280 \text{ ft}}\right) = 10.7° \text{ slope}$

FIGURE 25.1   Calculating slope using a topographic map

In this example, the units are English rather than metric because elevations on USGS maps are provided in English units and are intended for general public use. The formulas are the same for metric and English units.

name natural features, including mountains, valleys, floodplains, lakes, rivers, wetlands, and forests. They also identify human-made artifacts, such as roads, boundaries, transmission lines, and major buildings. Topographic maps are used for land-use planning, energy exploration, natural resources conservation, environmental management, public works design, and outdoor activities. In addition, historical topographic maps can be used to assess the growth and temporal impacts on natural resources.

AERIAL PHOTOGRAPHS	Aerial photographs are used to prepare topographical maps and are especially important in depicting land-use and land cover. The U.S. Geological Survey began using aerial photographs in the 1930s to construct topographical maps. Historical aerial photographs are also excellent sources for tracing land-use changes and identifying previous industrial operations, which can help identify potential areas of contamination. Most of the photos are black and white, but some are color-infrared, which are better for showing certain differences in plant life. Aerial photos are shot from airplanes flying at a constant altitude of 20,000 ft; each photo is shot straight down from the plane. Each 9"× 9" print covers an area of about 5 square miles at an approximate scale of 1:40,000, where an inch represents about 0.6 mi on the ground.

GOOGLE EARTH	Google Earth (http://earth.google.com) provides remote viewing of a site. It is a good source to determine adjacent resources and conditions because it allows you to see nearly anything, anywhere. Keep in mind though that the image may not be current.

SANBORN FIRE INSURANCE MAPS	Sanborn Fire Insurance Maps were published from the mid- to late 1800s to about 1980. They are a valuable resource for evaluating the potential for past contamination based on historical uses of urban property. Historically, some of the most serious fire risks were related to chemical, oil, and gas storage. Consequently, their location, size, and contents are noted prominently on historical fire insurance maps. In addition, insurance maps also note the name and/or type of manufacturing operations, which may indicate a potential contamination problem as shown in Table 25.1. Until 1980, waste management practices were generally substandard. Waste was dumped on the ground or in waterways, discharged through substandard sewers, or stored in temporary devices. Consequently, waste contaminants may have leached and contaminated the soil and/or groundwater.

The Sanborn Maps can be downloaded in GIF, JPEG, PEG200, or TIFF format from the Library of Congress (https://www.loc.gov/collections/sanborn-maps/about-this-collection/how-to-view-maps) for viewing. Many archival companies, local historical societies, and municipal resources have actual copies and digital copies.

Table 25.1 Manufacturing Activity and Likely Associated Chemical Pollutants

Manufacturing activity	Likely chemical pollutants
Leather tanning	Heavy metals, cyanide, chromium
Metal foundry	Heavy metals, solvents, oil, acids
Paint	Solvents, lead
Silverware	Heavy metals, cyanide, solvents, acids
Kerosene oil works	Oil, kerosene
Gas works	Oil, PNAHs (polynuclear aromatic hydrocarbons)
Photography	Mercury, silver, solvents, acids

Materials that come from these sites should have their recycling potential be carefully considered.

Your job is to determine the best place to site (locate) a new compost facility. The 35-acre compost facility will receive organic, compostable materials (i.e., food waste, compostable food service containers, animal carcasses, brush, leaves, grass, and nonrecycled paper products including paper towels, tissues, and napkins) from residential and commercial sources. Upon receipt at the compost facility, the organic materials will first be ground up prior to being composted in large windrows. Composting is a controlled decomposition process; during the process, the organic materials will continuously be aerated and monitored for temperature. During this time, the material should reach temperatures of over 180°F, effectively removing pathogens, weed seeds, insects, and residual insect larvae. Following the initial decomposition stage, the composted material will then be cured in large static piles for a minimum of 3 months. (The curing stage is carried out at a lower temperature, allowing for different organisms to further decompose the organic material.) Composted material will then be separated by type and then processed and sold as compost and mulch products.

The compost facility cannot be closer than 1 km to any surface waters because of potential runoff. The facility also will need to be located near access roads to reduce transportation costs, have access to electricity, and, because of potential odor problems, be located a sufficient distance away from residences.

1. Make a checklist that addresses each of the criteria below that are important in siting a compost facility.
   - Access roads
   - Critical habitat for endangered or threatened species
   - Depth to the groundwater
   - Distance, direction, and slope to the nearest population
   - Floodplains
   - Residential development
   - Slope
   - Suitability of the soil (no porous sand or gravel)
   - Surface water, wetlands, and natural resource features
   - Unstable areas (e.g., seismic areas, fault areas, karst topography)
   - Visibility

2. On graph paper, draw and then cut out the project's footprint (assume that the footprint of the project will be a square or a rectangle) using the appropriate scale for the USGS Quad map. See Figure 25.2 to get an idea of the composition of a compost facility.

3. Using the different maps, your checklist, and the project's footprint, identify a suitable location for the compost facility. List your preferred location and propose an additional location as an alternative site. (See Figure 25.3 for a conceptual application of the maps to your site project.) Next, develop a list with brief descriptions of other key factors that could affect your selection of your preferred location. These factors would be non-site-related factors such as socioeconomic, political, and cultural factors affecting host community acceptance.

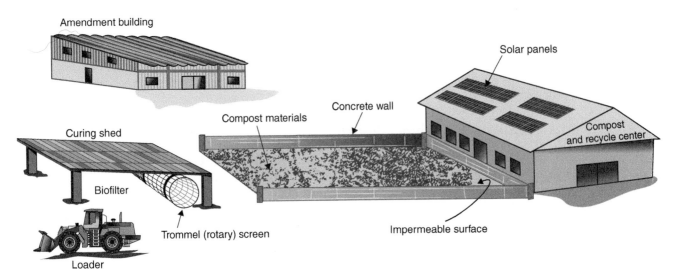

FIGURE 25.2 Sanitary compost facility

Sanitary compost facilities may be independent or co-located with landfills or recycling centers. The compost area will be located way from where it could result in groundwater or surface water contamination. The center may have biofilter layers and concrete staging and processing areas. The site may contain a place to store amendment materials for the compost, and to "cure" the compost. The finished compost may be processed for sale in bags or by the cubic yard.

| **WRITE-UP** | The write-up for this lab is a project evaluation. Basically, you will be writing a two- to three-page evaluation and recommendation of the proposed project using the following headings. Show all work and calculations in footnotes or endnotes to support your answers. |

Project Headings and Content:

1. **Title of Project**

2. **Proposed Compost Facility Location**
   - Identify your proposed and alternative locations. Provide the grid reference points for each of the four corners of your primary project location on the 7.5-foot USGS topographical maps. You should be able to tell from the scale that 70 mi is equal to 1 degree, 1 minute (60 minutes in a degree) is equal to 1.2 mi, and 1 second is equal to 0.02 mi. Locate the latitudinal and longitudinal reference lines on the map. Measure from the reference point to the project boundaries; remember to add or subtract as necessary.
   - Be sure to reference the proper USGS quad map name.
   - What is the scale of the quad map in English and metric units (1 cm on the map equals how many meters on the ground)?

3. **Site Characteristics** (for both sites)

   A. Land cover and current use in the area and on the proposed site

   B. Dominant soil types on the site

   C. Type, size, and distance of the nearest wetland

   D. Elevation (metric and English) of highest and lowest points (and the names) of your selected site

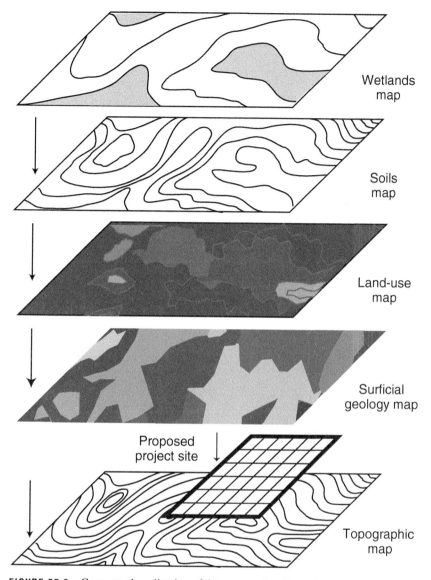

FIGURE 25.3 Conceptual application of the maps to the site project.

This type of overlay can be done with actual maps, with geographic information systems (GIS), or mentally.

E. Average slope of the project area (or slope characteristics if the area is many hectares)

F. Distance, direction, and slope (e.g., down- or upslope) to nearby natural resource features and the descriptions of these features

G. Distance and direction to the nearest settlement

H. Type, size, distance, direction, and slope to the closest water course (i.e., river, brook, or stream)

I. Type, size, distance, direction, and slope to the closest nonwetland surface water (i.e., lake, pond, or ocean)

4. **Community Acceptance**

   Describe the factors and likely impacts affecting the community's acceptance of hosting a new compost facility.

5. **Site Suitability**

   Based on your answers from #1 to #4, describe the overall suitability of your preferred and alterative site locations by highlighting strengths and weaknesses of the two locations.

REFERENCES

Library of Congress. 2017. Collections. Available at https://www.loc.gov/collections/sanborn-maps/about-this-collection/how-to-view-maps (verified 10 August 2017).

USDA NRCS. 2017. Soils Portal. Available at https://www.nrcs.usda.gov/wps/portal/nrcs/main/soils/survey/partnership/ncss/ (verified 10 August 2017).

Actual/Virtual Field Trip: Solid Waste Management

OBJECTIVES

- Describe how and where municipal solid waste is managed (i.e., landfills, incinerators, recycling centers, compost facilities, or waste-to-energy facilities) for a specific municipality.

- Estimate the amount of trash, recycling, and food waste generated by a local community.

KEY TERMS AND CONCEPTS

✓ Municipal solid waste

✓ Waste-to-energy

TIME REQUIRED: 2 to 3 hours for field trip option and 2 hours for virtual field trip option.

INTRODUCTION

Municipal solid waste, or MSW, (e.g., trash, refuse, or garbage) consists of materials we intend to discard and include everyday items including paper and cardboard, product packaging, plastic, grass clippings, furniture, clothing, glass bottles, food scraps, newspapers and magazines, and small appliances generated by households, businesses, and institutions such as schools. MSW can be categorized into trash, which will be buried (disposed of through landfilling) or burned (incinerated to capture the energy value), recyclables, (materials that can be recovered and recycled), and food waste (spoiled, uneaten portions or components of plated food), byproducts of food and beverage processing industries, and foods used to cook other foods (fats, oils, and greases) that can be composted or digested.

Historically, MSW management was a local responsibility. Most local governments continue to rely on landfills because, on a short-term basis, they are generally the least expensive disposal method (depicted in Figure 26.1). Landfilling, however, has potential significant long-term impacts including the generation of greenhouse gases (e.g., methane), groundwater and surface water contamination, and windblown litter. Modern incineration can represent a better alternative, but it is far more costly, especially when employing **waste-to-energy** (WTE), which uses combustion of the waste to produce electricity as a byproduct as shown in Figure 26.2. Both incineration and WTE reduce the volume of trash by 90%, but the resulting ash must be placed in a landfill. Of course, the best approach is to reduce the generation of MSW, followed by reuse and then by the recycling and the composting of food waste, which collectively can significantly decrease the need for disposal. The more we know about MSW and the proper roles of reduction, reuse, recycling, and composting, the better we can understand and reduce the impact of MSW.

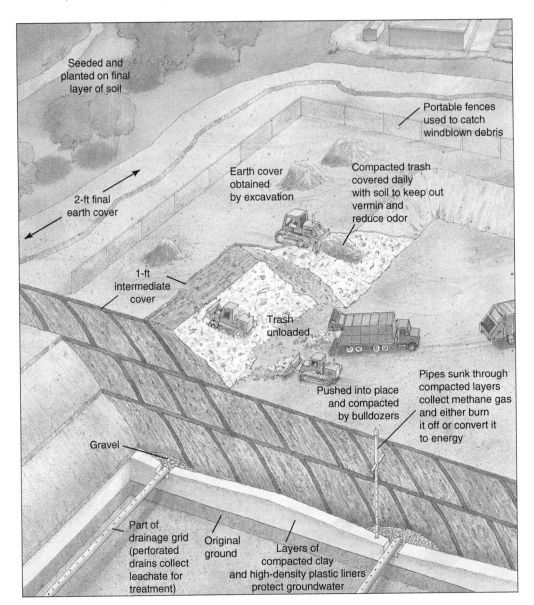

FIGURE 26.1 Municipal solid waste (MSW) landfill

Modern landfills are underlain with a monitored protective liner of high-density plastic or clay. Generally, above this layer, a leachate collection and removal system is installed to remove liquid leachate. Following each day of operation, the landfill is covered. Soil used to be the preferred cover, but recyclable materials and lightweight foams are now used.

Source: Raven, P.H., L.R. Berg, and D.M. Hassenzahl. 2008. *Environment.* 6th ed. John Wiley & Sons, New York (p. 560); and U.S. Environmental Protection Agency (US EPA). Landfill Methane Outreach Program [Online]. Available at http://www.epa.gov/lmop.

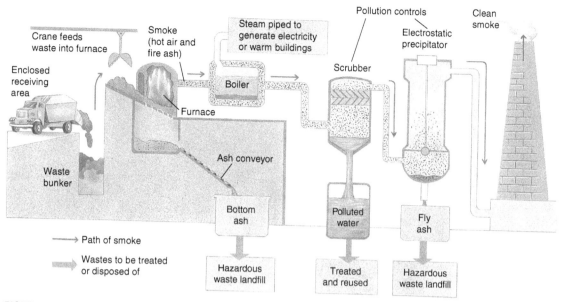

FIGURE 26.2   Mass burn, waste-to-energy incinerator

Scrubbers and electrostatic precipitators help trap noxious emissions. Fourth in the world, the United States burns 16% of its waste in waste-to-energy plants (1% is burned in old-fashioned incinerators that do not create energy). The number one waste-to-energy country is Japan (62%).

Source: Raven, P.H., L.R. Berg, and D.M. Hassenzahl. 2008. Environment. 6th ed. John Wiley & Sons, New York (p. 563); and Energy Information Administration (EIA). Annual Energy Review. Washington, DC. (See http://www.eia. doe.gov/ for current energy information.)

1. What is the name and location of the facility? (If it was a virtual tour, provide the URL for the tour.)

2. What type of facility is it (landfill, transfer station, composting, recycling, transfer station, etc.?)

3. When was it built? What is its capacity?

4. What geographic area does it serve? How many people live in this area? What is the per capita generation rate of MSW?

5. How much MSW does the facility handle/manage/dispose annually?

6. How has the type of MSW and recyclables changed over the past 10 years?

7. Are there seasonal differences in the amount of MSW generated?

8. What formal programs are in place to reduce the amount of MSW generated, capture recyclables, and/or compost food waste? If some of these programs do not exist, what are the biggest barriers to their adoption?

9. Is the recyclable materials market volatile (e.g., fluctuating prices for materials)? If so, why?

10. What is the greatest current challenge facing the facility?

11. What is the likely greatest future challenge facing the facility?

12. What are the future plans of the facility?

13. What is the life expectancy of the facility? At the end of its operating life, what are the plans to handle the MSW?

WRITE-UP

The write-up for this lab is a technical report. Use the following headings for your report. The tone of your report should be neutral and fact-based.

Title:

I. **Introduction** (name and location of the facility, geographic area served by, year built, capacity, and date and time of field or virtual site visit and the URL)

II. **Operation of the Facility** (describe the basic processes in managing MSW and also to reduce environmental impacts of the operation such as protecting groundwater, reducing stormwater runoff, reducing windblown litter, air emissions, and odors)

III. **Operational Challenges** (describe the current and future challenges facing the facility)

IV. **References Cited**

Testing the Toxicity of Chemicals

OBJECTIVES

- Understand how to test a material's toxicity.

- Count and categorize organisms using a dissecting microscope.

- Create a concentration–response curve for the toxicity of a material.

KEY CONCEPTS AND TERMS

✓ Acute toxicity

✓ Chronic toxicity

✓ Concentration–Response Curve

✓ Lethal concentration 50% (LC_{50})

✓ Toxicity

TIME REQUIRED: 2 to 3 hours.

INTRODUCTION

How do government and industry determine acceptable levels of human exposure to pesticides, new chemical products, cosmetics, and household chemicals? A major method is through toxicity testing to generate a concentration–response curve. In this lab, you will become familiar with the methodology of generating a concentration–response curve and investigate the advantages and disadvantages of this approach. A **concentration–response curve** (a variation of the dose–response curve depicted in Figure 27.1) is used to help assess the effects of various concentrations of a chemical substance on a group of organisms. Notice that the term "chemical substances" is used instead of "toxic substances." Every chemical is toxic; it is the dose, or concentration, that makes the poison. For example, table salt will cause an adverse effect in humans, but it takes a relatively large dose. In contrast, only very small doses of cyanide are necessary to cause an adverse effect. In this lab, you will be assessing **acute toxicity**, which is the adverse effects of a substance from a single exposure or from multiple exposures in a short period of time (generally less than 24 hours). In contrast, **chronic toxicity** is the adverse effects of a substance from long-term exposure.

To estimate toxicity, a substance is generally tested at various concentrations (or doses) often using living test organisms to determine what concentrations elicit a response in organisms. Selection of a test depends on many factors, including the potential target organ, whether it is a **chronic** test (long term, such as cancer or liver disease) or an **acute** test (short term, such as poisons), and available resources (Figure 27.2). Chronic toxicity testing, such as the testing of carcinogens, can be very expensive.

One standard measurement of acute toxicity of a substance is the lethal concentration 50% (LC_{50})—the concentration that causes death to 50% of test organisms. The lower the LC_{50} value for a substance, the higher the toxicity.

Measuring toxicity requires a toxicologist to plot data in the form of a concentration–response curve. This curve relates the concentration of a chemical to the percentage of animals showing the response (e.g., death). This curve will allow you to determine the concentration of a toxic material that causes 50% mortality in a population of test organisms. Although the goal is to

(a) General response curve

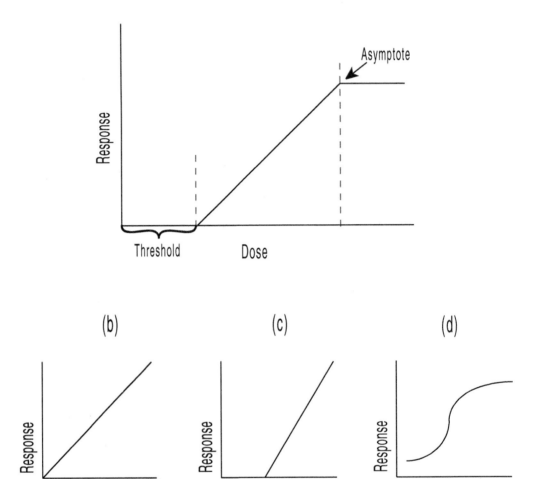

FIGURE 27.1 Generalized response curve

(a) An example of a general response curve. The *threshold level* is the level at which an adverse effect is not observed. The *asymptote* represents the maximum effect (response) level. (b) An example of a *linear response curve* where a unit increase in the dose produces a unit increase in the response. This curve has no threshold level. Current national policy assumes that there is no "safe" dose of a carcinogen; thus, there is no threshold. (c) An example of a *linear response curve* with a threshold level. Thus, there is a certain dose where no adverse effect is observed. (d) An example of a *nonlinear response curve*, in which the shape of the relationship is curved. Thus, a unit increase in dose will have varying responses. Some nonlinear response curves have a threshold and/or an asymptote—when the dose reaches an asymptote, no additional dose affects the response.

establish the 50% level, knowing that the shape of the curve above and below the midpoint is equally important. A hypothetical example of a concentration–response curve is depicted in Figure 27.3, but note that the response might be much more linear or it might have a lower threshold.

The response to the concentration is simply the amount of damage it causes. There may be visible or measurable symptoms of toxicity at sublethal levels. A common endpoint for

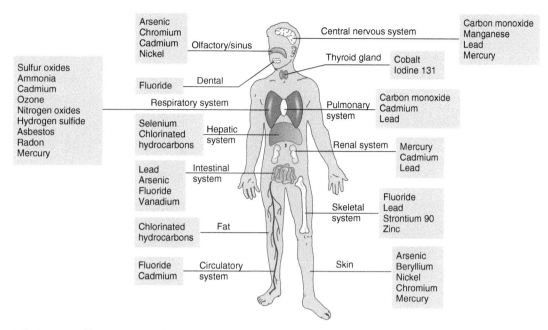

FIGURE 27.2   Target organs and pollutants

The effects of some major pollutants in humans.

Source: Botkin, D.B., and E.A. Keller. 2007. Environmental Science: Earth as a Living Planet. 6th ed. John Wiley & Sons, New York (p. 307); and Waldbott, G.L. 1978. Health Effects of Environmental Pollutants. 2nd ed. Moseby, St. Louis, MO.

most chemicals tested on organisms is death because it is easily identifiable, reliable, and cheaper than sublethal assessments.

Common test organisms include zooplankton, algae, mice, and rats (rats and mice account for 90% of the animals used in toxicity testing); the results are often then extrapolated to humans, combined with safety and uncertainty factors, to estimate an acceptable exposure level. Because of our vastly increased knowledge of toxicity and the ethical concerns with testing on animals, the reliance on animal testing has been reduced.

In this lab, you will perform an experiment to examine the effects of a substance on an organism and calculate a simple concentration–response curve.

MATERIALS

- Beakers
- Chemical (e.g., household pesticide, vinegar, ammonia, cleaning compound)
- Dissecting microscopes
- Graduated cylinders
- Labels
- Lower-order organisms, 100 individuals (e.g., brine shrimp, nematodes, Daphnia)
- Petri dishes, five per team (either square dishes with etched grids or round dishes underlain by graph paper)
- Pipettes

Concentration–response curve of a chemical

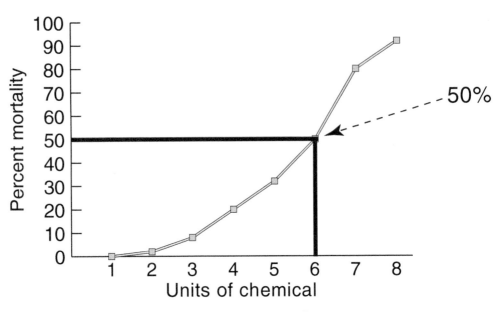

FIGURE 27.3   A hypothetical concentration–response curve LC$_{50}$

The curve may be quite different depending on the substance and the measurements. Compare it with the dose–response curves shown in Figure 23.1.

TASKS

1. Based on your chemical, go to the library to conduct background research (including the chemical's Material Safety Data Sheet) to help identify what a possible lethal concentration could be. Use what you found help formulate a hypothesis regarding the potential lethality of your substance. Remember that you are hypothesizing what the LC$_{50}$ is (x% concentration of your contaminant will kill 50% of the organisms).

2. Set up the experiment:
   • Obtain six petri dishes. Place 15 mL of solution appropriate for the organism into each petri dish. (e.g., if using brine shrimp, use a prepared NaCl solution to maintain the shrimp.)
   • Select five different concentrations to test. (Use Table 27.1 as a guide.). Select your treatments so that they bracket the effect you have predicted. That is, because your goal is to find the LC$_{50}$, select two concentrations above and two below your hypothesized concentration. (One of the petri dishes is your control in which no chemical will be added). For example, following Table 27.1, if you predicted LC$_{50}$ is 11.76%, choose 3.22% and 6.25% as the two concentrations below and then 16.67% and 21.05% as the two concentrations above. (We are using two decimal places here and in Table 27.1 but it is reasonable to round to one decimal place.)
   • Label each petri dish by concentration.
   • Add the required amount of concentration into each petri dish. Gently swirl, mixing thoroughly to ensure that the concentration is the same throughout the petri dish.
   • Carefully place 20 live organisms into each petri dish and immediately record the time.

Table 27.1 Some Contaminant Concentrations for Toxicity Testing

Dish	Solution (mL)	Contaminant (mL)	†Calculate	Contaminant concentration
*C	15	0	0	0.00%
1	15	0.5	0.5/15.5	3.22%
2	15	1	1/16	6.25%
3	15	2	2/17	11.76%
4	15	3	3/18	16.67%
5	15	4	4/19	21.05%

*This is your control.

†To obtain an accurate concentration, you must also add the volume of the contaminant to the volume of the solution.

- Wait 5 minutes, count and record the number of dead into Table 27.2. Then recheck in 5 more minutes; this will be your final mortality count. (Note that your results must be labeled as 10-minute LC_{50} as most LC_{50} tests are for multiple hours.) Determine the percentage of organisms that died in each petri dish.

3. Based on your initial results, did you accept or reject your hypothesis? That is, did you find the LC_{50}?

4. Construct a dose–response curve following the Excel instructions listed below.

5. Based on your initial results, repeat the experiment but use slightly different concentrations to produce a more robust concentration–response curve as you will now have twice as many as data points. Follow the procedures in #2 above, and record your data into Table 27.2.

6. Construct a single concentration–response curve with all data points similar to Figure 27.4. The dose–response curve can be created in Excel as discussed in the textbox.

Table 27.2 Data Collection Table for Toxicity Testing

	Concentration	5-minute LC_{50}	10-minute LC_{50}	Percent mortality
C	Control			
1				
2				
3				
4				
5				
C	Control			
6				
7				
8				
9				
10				

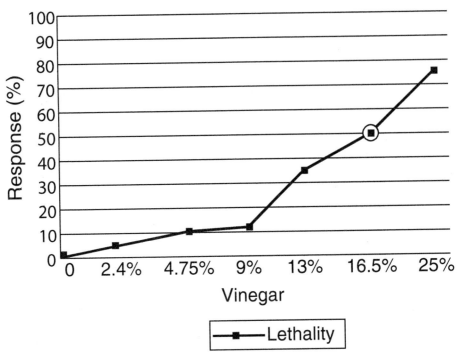

FIGURE 27.4 The LC_{50} for vinegar on a hypothetical test organism

In this hypothetical case, the LC_{50} for vinegar is 16.5%.

<table>
<tr><td>**WRITE-UP**</td><td>A formal experiment laboratory write-up is required.</td></tr>
</table>

Title:

I. **Introduction**: Discuss the issues of toxicity and human health. Discuss your household chemical. What does the background literature say about toxicity for your chemical and for your test organism? End this section by stating your hypothesis (e.g., your predicted LC_{50} for your chosen chemical).

II. **Methods**: How did you conduct the test and collect your data?

III. **Results**: What did you find, be sure to include your concentration–response curve.

IV. **Discussion**
- Discuss whether you accept or reject your hypothesis.
- Based on your results, classify your contaminant as Very Toxic, Relatively Harmless, and so forth. For this, you will have to conduct Internet research (see, for example, the Hodge and Sterner Scale) to find a source that correlates an LC_{50} level with qualitative descriptors. Cite this source in your References Cited section.

- Describe the potential areas of weaknesses in your conclusions (e.g., sampling, extrapolation, assumptions).
- Briefly describe what you would do next time to improve the results.
- Address the difficulties in extrapolating your lab data to real-world situations (i.e., using your data to determine a safe level for human exposure).

V. References Cited

Instructions for Generating a Concentration–Response Curve in Microsoft Excel

1. Open up Excel. In the first row, input the contaminant concentrations that you tested.

	A	B	C	D	E	F	G	H
1	Mortality	2	8	12	18	26	48	64
2	Concentration	2.5	5	7.5	10	12.5	15	20

2. In the second row, under each concentration, input the percentage of organisms that died at that concentration (mortality). Continue until all percent concentrations and mortalities have been input.

   (Remember that this is a concentration–response curve; the curve must be flat or increase, it cannot decrease.)

3. After inputting the data, highlight only the Mortality row and then left-click Insert. Then left-click Line under Charts. There should be a single line only.

4. Right-click the Line graph, click on Select Data. On the right side, under Horizontal (Category) Axis Labels, left-click Edit. In the box titled Axis label range, highlight the concentration row cells (B2 through H2 in the above example), but do not highlight cell B1 containing the label Concentration. Select OK and then select OK again and the concentrations should now be on the X axis.

5. In the Layout tab, you can now add labels to the figure and to the X and Y axes.

6. After you have completed the chart, you can copy the entire chart with a right-click and then paste into a Word document, PowerPoint presentation, or other software for your report.

Environmental Risk Perception

OBJECTIVES

- Understand how individuals' perceptions of risk can differ.

- Collect data to rank environmental issues in terms of risk.

- Explain variations in risk perceptions.

KEY CONCEPTS AND TERMS

✓ Environmental risk

✓ Exposure

✓ Risk communication

✓ Risk perception

✓ Outrage factors

TIME REQUIRED: 2 hours.

INTRODUCTION

Environmental risk refers to the potential of an agent (chemical or physical) to cause harm to human health and/or the environment. We constantly face environmental risks from the air we breathe, water we drink, soil and water we touch, and food we eat. For a harm to occur, there must be **exposure** (the potential for a person to come into contact—ingestion, inhalation, direct contact—with an agent, such as a pollutant). Some risks are more significant than others and depend on a multitude of factors, including the toxicity and concentration of a contaminant, the number of people exposed, the short- and long-term frequency of exposure, where we live, our age, our body weight, our daily activities, and contributing factors (e.g., smoking, diet, alcohol). For example, indoor air quality can present a much higher risk, in part, because of the high exposure potential as we spend far more time indoors than outdoors. And a child exposed to the same concentration of a contaminant as an adult is at a much greater risk because there can be a higher dose in relation to the child's body weight.

A major factor in how and what environmental risks we manage depend on how we perceive risks. This factor—**risk perception**—is an individual or group assessment based on a feeling, belief, or judgment of the potential for an environmental factor or condition to cause harm. For example, most people perceive the risk from hazardous waste to be very high, but most environmental experts rank the risk to be relatively low. In contrast, most people rank the risk of contaminated groundwater—a major source of drinking water—from gasoline stored in leaking underground storage tanks to be low, whereas many environmental experts rank this risk much higher. A major reason for this difference is a person's perception, which is based on education, ethnic background, familiarity, past experiences, and a variety of socioeconomic and cultural factors. Risk perception is also heavily influenced by how the media reports on a risk because the media can sensationalize or downplay a risk such that it can skew one's perceptions.

MAJOR OUTRAGE FACTORS

Voluntariness: Degree to which we choose to be exposed.

Control: The degree to which we control the activity compared to another person or organization.

Fairness: Shouldering more of the risk without benefits compared to others.

Trust: The degree to which we trust the responsible organization.

Familiarity: How familiar we are with the technology or activity.

Memorability: The vividness, the ability to recall an accident or tragic event.

Dread: How much we dread the harm such as cancer.

Catastrophic potential: If an event occurs, can it harm a significant number of people at the same time, such as an airplane accident?

Risk communication is the purposeful exchange of information about risks among interested parties. A key component of risk communication is understanding how experts and the general public may differ in the communication, understanding, and perception of risks. According to Sandman (1987), risk experts focus on the quantitatively based potential for harm while the general public focuses on what he terms **outrage factors**, which are influential factors that shape our individual perception of environmental risks.

No special materials are needed for this lab. However, before engaging in any survey, contact your campus Institutional Review Board to determine any applicable requirements and/or approvals.

MATERIALS

In this lab, you are going to conduct research using a survey to gather data on people's perception of environmental risks.

TASKS

1. Below is a list of environmental issues. As a team or a class, select 10 environmental issues from the list (or include your own based on what is the most current). You can list these risks using the suggested survey template—be sure to list them in alphabetical or random order. Each team will need at least 20 copies of the survey forms.

Environmental issues	
Genetically modified organisms	Food additives
Habitat loss	Solid waste
Acid rain	Invasive species
Air pollution	Stormwater runoff
Oil spills	Lawn-care pesticides
Asbestos	Underground storage tanks
Ozone depletion	†Ground-level ozone
*Climate change	†Smog
*Global warming	Hazardous waste
Water pollution	Urban sprawl
Radon	Indoor air pollution
Drinking water quality	Litter

*,†: See the box "Importance of Wording."

2. In your team, hypothesize what you believe the results will be by ranking the issues based on their risk, from 1 to 10. The issue that presents the greatest risk would be number 1 and then rank the other issues sequentially down to number 10. You are predicting what your results will be.

3. In teams, using the survey sheets, randomly select *at least 20 individuals* to complete the survey. Unless otherwise instructed, go to a variety of locations and do not limit yourself to your building or just students.

4. Ask each participant to rank the listed environmental issues using an ordinal rank system, based on their perception of the risk, by sequentially ranking each issue from 1 to 10 by

---

IMPORTANCE OF WORDING

Note that some of the environmental issues appear to be similar but can be perceived quite differently (e.g., climate change and global warming, ground-level ozone and smog). For example, research has shown that survey respondents perceive global warming as being primarily human-caused and involve primarily only warming. In contrast, climate change is viewed primarily as more natural and thus less intervention is needed. Wording is one of the greatest challenges of a valid survey. As an experiment, try using different terms for the same environmental issue to see if participants distinguish between the two.

---

(1) presenting the greatest risk of the list of issues and (2) presenting the lowest risk of the list. Each issue will have its own rank in relation to each other issue.

5. It is imperative that you avoid answering specific questions on any of the environmental issues. By answering questions, you will be influencing or educating the participant and thus skewing your results.

6. After completing the surveys, return to the lab and tally the results. (If this was a class effort, your results will be tallied as a class otherwise tally your team's results.) Use percentages and means to interpret your results. In your write-up, answer the following questions:

A. What was viewed as the highest risk? Are there any discernible patterns (e.g., students vs. faculty, source of news)?

B. What was viewed as the lowest risk? Are there any discernible patterns (e.g., students vs. faculty, source of news, etc.)?

C. To what degree did any of the **outrage factors** play a role in individuals' perception of risk?

D. What surprise findings resulted from the survey?

E. How accurate are your results? Are the results an accurate representation of your town, city, or state based on sample population, sample randomness, sample size, and time and location of sampling?

F. If you were a public official for your municipality, based on the results of your survey, how would you respond to the following:
   I. What environmental issues would you focus on and why?
   II. As a public official, what other factors should be considered besides the environmental risk?

| WRITE-UP | Your lab report will be a technical report. Although you may have worked as a team, be sure to write and submit your own report unless instructed otherwise. Use the following format: |

Title:

I. **Introduction:** Background information explaining environmental risks. End this section regarding your hypothesized ranking results.

II. **Methods:** Explain who, what, and how you collected your data through surveying.

III. **Results:** Present your findings; be sure to use a table or graph.

IV. **Discussion:** Interpret your results, using your responses to 6.A to E. Be sure to discuss your responses to 6.F.

| REFERENCES | Sandman, P. M. 1987. Risk Communication: Facing Public Outrage. EPA Journal. 13: 21–22. |

APPENDIX

Environmental Risk Survey

Hello. We are conducting an anonymous survey for a school project. Completion of the survey will take approximately 4 minutes. We will keep your answers confidential and you will not be identified as a respondent.

Below is a list of environmental issues presented in alphabetical order. Please rank them in order from 1 to 10 based on the degree of environmental risk each presents: 1 being the highest environmental risk among the list and 10 being the least risk.

Environmental issue	Rank

Please complete the information below as it applies to you:

Gender: _____Male _____Female _____ Other

Status: _____Part-time student_____Full-time student_____Staff_____Professor _____Visitor

Primary source of news: Television _____ Facebook _____ Twitter _____

Newspaper _____ Other _____

Thank you for your participation.

Human Survivorship Changes

OBJECTIVES

- Describe how human mortality and survivorship have changed in the past 200 years.

- Collect population data and generate mortality and survivorship graphs with Excel.

KEY CONCEPTS AND TERMS

✓ Demography

✓ Population

✓ Survivorship curve

✓ Type I survivorship

✓ Type II survivorship

✓ Type III survivorship

TIME REQUIRED: 2 hours of field time if visiting a local cemetery, plus 1 to 2 hours for online research.

INTRODUCTION

A key factor in environmental science is **population**—a group of individuals of the same species living in the same area. **Demography**, the study of changes in population for any species, is based on a simple model: (Births – Deaths) + (Immigration – Emigration). If these rates remain equal, the population will generally remain the same. However, if any of the inputs or outputs changes due to internal or external factors, the population is affected. For human populations, there are many such examples throughout history: the high death rate from the bubonic plague in 14th-century Europe, the mass emigration of the Irish during the 19th-century potato famine, World War I (1914 to 1918) combined with the influenza pandemic of 1918, and the post-World War II baby boom in America. Ecologists generally consider humans to exhibit **Type I survivorship** because the probability of survivorship is fairly high at birth; **Type II survivorship** is more linear (examples include some species of lizards). **Type III survivorship**, typical of many fish, has high initial mortality (see Figure 29.1). Within different human populations, there can be a range of survivorship patterns.

An important question for humans is the impact of longevity on population. That is, how does life expectancy fit into our population model? Americans in the 21st century, on average, have a much longer life expectancy than they had in the 19th century. Among the factors that affect longevity are better health care, improved nutrition, better working conditions, higher education, and new technologies. According to Botkin and Keller (2007), the life expectancy in ancient Rome was about 30 years.[1] In some countries, life expectancy has not improved much since then; it is only 39 in Botswana and some other African nations. However, in America, average life expectancy is now around 78 (Arias, 2014). If we are living longer, does that necessarily mean a higher population growth? This can be a complicated question because it depends on birth rate (e.g., do women with a long life expectancy have more children than do women with short life expectancy?) and other environmental, cultural, and socioeconomic factors. An important environmental issue is whether the consumption and use of resources to support longer lives

---

[1] There is some debate over the actual number of years, and some researchers think that Roman life expectancy was even lower. A good discussion can be found in Parkin (1992).

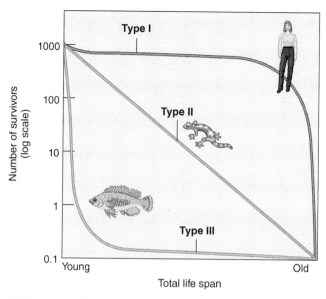

FIGURE 29.1   Survivorship types

These generalized survivorship curves represent the ideal survivorships of species in which death is greatest in old age (type I), spread evenly across all age groups (type II), and greatest among the young (type III).

Source: From Raven, P.H., L.R. Berg, and D.M. Hassenzahl. 2010. Environment, 7th ed. John Wiley & Sons, New York. (p. 177).

means that there is less for the rest of the population. This issue is important when considering human population, available resources, sustainability, human rights, and public policy.

These are interesting and important considerations, but we need basic data before we can assess the potential implications. Accordingly, the primary purpose of this lab is to calculate and compare life expectancies with a **survivorship curve**, which is a graphical representation of the likelihood that an individual will survive from birth to a particular age (see Figure 29.2).

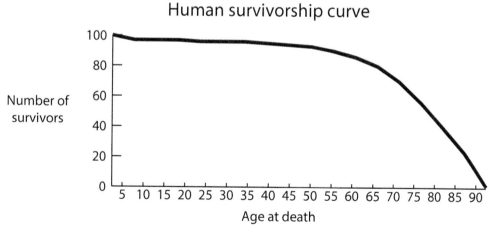

FIGURE 29.2   Sample human survivorship curve

Assuming a population of 100, this figure depicts the number of individuals alive at each year of age—survivorship. Note that after age 65, the number of survivors declines rapidly. If this population sample had experienced a major war or epidemic, the results might show up as one or more dips in the curve before the end of normal life expectancy. Overall, the human survivorship curve is Type I.

Survivorship curves are constructed using gender and birth and death data. By comparing survivorship curves for different periods of time, you can identify historical trends in population demographics over any time period. Based on this comparison, you can speculate on why there was a change and the ramifications for population and the potential environmental impact. Human birth and death data can be obtained from gravestones, city/county records, and newspaper obituaries. In this version of a commonly practiced demographics exercise, you will collect data and construct comparative survivorship curves.

MATERIALS   No special materials are required.

TASKS

1. Create a hypothesis as to which group (pre-1900 females, pre-1900 males, post-2000 females, or post-2000 males) has the longest longevity.

2. Take a trip to a nearby cemetery to gather data on life span for a sample of people in the 19th century (or use the data provided below). You will also need to go to a library or use the Internet to gather data for a more current sample. A large cemetery may have sufficient resources to accommodate the "old" and "modern" sample collection. Use the following sequence:
   - For the "old" sample, examine headstones or old newspaper obituaries (or use the data below) in a nearby community and record the age at death for males *and* females (at least 25 each) who died prior to 1900. Do this during lab time. Each person is to gather his or her own data.
   - For the "modern" sample, obtain the age at death for males *and* females (at least 25 each) in the community who died within the past 5 years. You can get this information from obituary pages online. Try to obtain the data for the same community for both samples.
   - Enter the data in an Excel spreadsheet and construct four separate curves (pre-1900 females, pre-1900 males, post-2000 males, and post-2000 females) on one graph. Follow the instructions below for constructing the curves. Be sure to put death age on the X axis and number surviving on the Y axis. This should be straightforward if you are using the data in spreadsheet format. Remember that you are graphing *survival* curves, not *death* curves; thus, your curve can only go down, it can never go up! Also use appropriate dimensions and labels for your graphs, including the X and Y axes and the title.

3. Calculate the mean longevity per group.

4. Comment on the different curves. What do they suggest to you about your samples?

5. Identify and describe the likely environmental, cultural, and socioeconomic factors that make the survivorship curves different.

6. Discuss the environmental implications of longevity in modern societies by focusing on a particular topic such as fossil-fuel-derived energy, climate change, biodiversity, water consumption, or availability of seafood.

Option—Hypothetical data for constructing survivorship curves			
Pre-1900 males, age of death	Pre-1900 females, age of death	Post-2000 males, age of death	Post-2000 females, age of death
16	6	67	87
23	18	79	72
34	49	66	86
62	71	62	62
1	57	83	17
3	16	57	83
57	1	68	83
34	49	94	68
57	58	73	94
44	43	21	73
49	39	68	57
36	19	74	68
24	42	18	94
18	18	1	48
2	76	19	71
39	59	72	69
37	61	73	61
51	64	63	73
48	1	76	78
60	61	64	77
72	69	71	72
69	72	69	87
53	54	84	79
59	67	87	62
71	63	79	84

	A	B	C	D
	Pre-1900 females	Pre-1900 males	Post-2000 females	Post-2000 males
1	25			
2	23			
3	23			
4	23			
5	23			
6	22			
7	22			
Σ	etc.			

INSTRUCTIONS TO PRODUCE SURVIVORSHIP GRAPHS IN EXCEL

INSTRUCTIONS FOR GENERATING A SURVIVORSHIP CURVE
IN MICROSOFT EXCEL

1. Open up Excel. Choose your first category (e.g., pre-1900 females), and in the first column (A) and first cell (1), input the title of the category. Next, in Column 1, input the number of individuals alive on their first birthday (e.g., at age 1, 25 individuals were alive). Continue to do this until all individuals have been input. For example, at age 2, 23 individuals were alive. At age 13, 22 individuals were alive and at age 78, only 1 person was alive (at age 79, 0 persons were still alive).
    Remember, this is a survivorship curve; the curve must always be decreasing.

2. Put your next category (e.g., pre-1900 males) in Column B and input your data on number of individuals alive at each age, then input the remaining data in Column C and Column D.

3. After inputting the data, highlight all the data and then left-click Insert. Then left-click Line under Charts. There should be four individual lines all sloping downward.

4. In the Layout tab (or add Chart Elements tab), add labels to the figure and to the X and Y axes.

    After you have completed the chart, you can copy the entire chart with a right-click and then paste into a Word document, PowerPoint presentation, or other software for your report.

WRITE-UP

Your lab report will be a technical report. Use the following format:

Title:

I. **Introduction:** Background information explaining human survivorship. End this section regarding your hypothesized expectations of the differences you might find.

II. **Methods:** Explain how and where you obtained your data.

III. **Results:** Present your findings including your survivorship curves.

IV. **Discussion:** Interpret your results, using your responses to the task questions and be sure to relate your answers to your own experience.

V. **References Cited:** Be sure to cite any references, including online sources of obituaries or vital statistics used.

REFERENCES

Arias, E. 2014. United States Life Tables, 2010. National Vital Statistics Reports 63(7). National Center for Health Statistics, Hyattsville, MD. Available at https://www.cdc.gov/nchs/data/nvsr/nvsr63/nvsr63_07.pdf (verified 15 August 2017).

Botkin, D. B., and E. A. Keller, 2007. Environmental Science: Earth as a Living Planet. 6th ed. John Wiley & Sons, New York.

Parkin, T.G. 1992. Demography and Roman Society. The Johns Hopkins Press, Baltimore, MD.

Indoor Air Quality Inspection

OBJECTIVES

- Describe the major sources of indoor air pollution.

- Measure carbon dioxide levels, humidity, temperature, and flow in the air.

- Describe how indoor air pollution problems are addressed.

KEY CONCEPTS AND TERMS

✓ Air exchange rate

✓ Indoor air quality

✓ Infiltration

✓ Mechanical ventilation

✓ Natural ventilation

✓ Relative humidity

✓ Sick building syndrome

TIME REQUIRED: 2 hours.

INTRODUCTION

In the early 1970s, the global oil shortage prompted a national focus on conserving energy. One particular area of focus was reduction in the loss of heat and cooling due to poor insulation and ill-fitting windows in buildings. At the same time, with advancements in materials science, industry was increasingly manufacturing household products and building materials using advanced plastics, solvents, glues, and other chemicals (e.g., formaldehyde), as shown in Figure 30.1. By "tightening" buildings (i.e., eliminating air leaks to reduce heating and cooling losses) and simultaneously increasing the use of potentially hazardous materials indoors, we created a potential exposure problem because most Americans spend up to 90% of their time indoors. This exposure is significant because adult lungs intake about 16,000 qt of air each day. Studies conducted by the US EPA and others show that indoor environments sometimes can have levels of pollutants that are actually higher than levels found outside (US EPA, 1998).

Ambient (outside) air enters and leaves a building or other structure through infiltration, natural ventilation, and mechanical ventilation. With **infiltration**, outdoor air flows in structures through openings, joints, and cracks in walls, floors, and ceilings, and around windows and doors. In **natural ventilation**, air moves through opened windows, doors, and passive vents. **Mechanical ventilation** devices, such as outdoor-vented fans, are used to intermittently remove air from a room to air-handling systems that distribute filtered and conditioned outdoor air to strategic points throughout the building. The rate at which outdoor air replaces indoor air is defined as the **air exchange rate**. When there is little infiltration, natural ventilation, or mechanical ventilation, the air exchange rate is low, and pollutants can build up, including carbon dioxide from respiration.

As the flow of fresh air into buildings is being reduced and indoor pollutants increased (including second-hand tobacco smoke), indoor air quality decreases. This leads to **sick building syndrome**—a set of symptoms that affect some occupants during time spent in a building and diminish or disappear when away from the building, but cannot be traced to specific pollutants or sources within the building. An **indoor air quality** problem exists when four elements

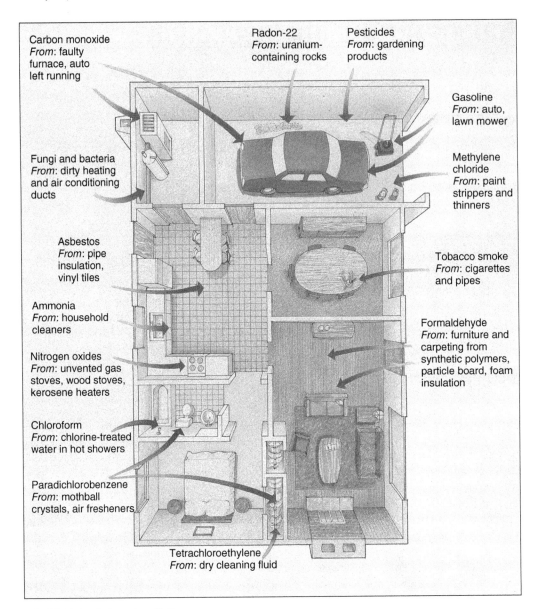

Carbon monoxide
From: faulty furnace, auto left running

Radon-22
From: uranium-containing rocks

Pesticides
From: gardening products

Gasoline
From: auto, lawn mower

Fungi and bacteria
From: dirty heating and air conditioning ducts

Methylene chloride
From: paint strippers and thinners

Asbestos
From: pipe insulation, vinyl tiles

Tobacco smoke
From: cigarettes and pipes

Ammonia
From: household cleaners

Formaldehyde
From: furniture and carpeting from synthetic polymers, particle board, foam insulation

Nitrogen oxides
From: unvented gas stoves, wood stoves, kerosene heaters

Chloroform
From: chlorine-treated water in hot showers

Paradichlorobenzene
From: mothball crystals, air fresheners

Tetrachloroethylene
From: dry cleaning fluid

FIGURE 30.1 Indoor air pollution

Air quality inside a building cannot be any better than the air outside, but good air ventilation can make a big difference.

Source: Raven, P.H., L.R. Berg, and D.M. Hassenzahl. 2008. Environment. 6th ed. John Wiley & Sons, New York (p. 478).

are present: (1) a contaminant, (2) an exposure pathway to a person, (3) a person (also called a receptor), and (4) a driving force to move the contaminant. If any of these elements is removed, the problem can be eliminated (or at least postponed). For example, a chemical leak in an unoccupied warehouse is not a direct indoor air quality problem for humans (no exposure means no human health effect). Although there is a source, a pathway (the air in the building), and a driving force (drafts and ventilation that circulate air), if the building is unoccupied, there is no human exposure. In comparison, if there is mold contamination in the ductwork of an occupied office building, there is a source of contamination (mold), a pathway (the duct work), a driving force (the air flowing through the duct), and human exposure (occupants of the building).

In addition to oxygen and other gases we need to survive, air contains particulate matter, volatile organic compounds, biologicals, and water vapor. **Relative humidity** is the percentage of water in the air compared to the maximum amount of water that the air will hold at that temperature. The amount of moisture in the air is limited by temperature and pressure. Moisture content affects the weight density, which influences how air flows. Air has a weight density of 75 lb per $1,000 ft^3$ ($1.201 kg/m^3$) at room temperature. Water, by comparison, weighs 830 times as much as air. Air has a composite molecular weight (MW) of about 29 g/mol.

Standard temperature and pressure (STP) is assumed to be 20°C (68°F) and 760 mmHg (29.92 in.Hg).[1] Pressure differences cause air to move. Therefore, for a contaminant to move from a source to a person, there must be a change in pressure. Atmospheric pressure pushes air from a high-pressure environment to a lower-pressure environment. In buildings, air is moved in ducts and a fan is used to create a pressure difference. A fan lowers atmospheric pressure (creates negative pressure), and atmospheric pressure moves air into the duct in order to equalize the pressure. Negative pressure wants to collapse the duct and positive pressure wants to expand it. These pressures are used in buildings to control temperature, airflow, and contaminants. For example, reactor containment buildings at nuclear power plants are under negative pressure; a small radioactive leak would tend to stay inside the containment building. To ensure this condition, less air is supplied to the containment building than to the surrounding environment. Air from the surrounding area will be pulled into the building to compensate for the negative pressure differential, thereby eliminating the possibility that radioactive gases can leak out. An example closer to home might be that of a hotel with an indoor swimming pool. The indoor pool should be in a negative pressure enclosure because it prevents the chlorine smell from entering the lobby and hallways. You can tell right away if that pool is not under negative pressure.

In this lab, you will investigate the indoor air quality of a specific building using simple, common techniques and will be testing for the following:

- Biological contamination

- Carbon dioxide levels

- Humidity

- Pressure and ventilation as indicated by airflow

- Temperature

MATERIALS

- Calculator

- Carbon dioxide (CO_2) meter

- Digital thermometer

- Hygrometer

- Soap bubble solution and wand

TASKS

You will be conducting a site investigation into the indoor air quality of a specific building. While sophisticated equipment exists to measure the exact pressure in ducts and building supply and exhaust vents, you will use a very simple technique to determine pressure differentials in a building.

[1] Correction factors are used when air is measured at other than standard conditions to account for the change in air density.

DETERMINING PRESSURES

- In your room, close the door.
- About 1 ft away from the door in the inside of the room, gently release soap bubbles near any gaps (e.g., above the gap between the door and the floor).
- Describe what happens to the bubbles (if they move away from the door there is negative pressure inside the classroom but toward the door there is positive pressure in the classroom).

VENTILATION SYSTEMS

Ventilation systems are designed to support the maximum occupancy of a space. When we breathe, we inhale oxygen and exhale CO_2. If the building's ventilation system is supplying and exhausting an appropriate amount of air to a space, levels of exhaled CO_2 will not build up in a room. (Sustained levels of CO_2 above 1,000 ppm usually indicate inadequate ventilation.) Accordingly, CO_2 can be used as an indicator of appropriate ventilation. During an initial indoor air quality investigation, CO_2 levels are measured throughout the day and analyzed to determine if the ventilation system is working correctly.

---

1. Test the ambient air—the air outside of the building—for relative humidity, temperature, and CO_2. Input your data into Table 30.1.

2. Select an empty classroom that has not been occupied for at least 4 hours. (Be sure that the room has a door that can be closed.)
   A. Inside the empty classroom, test for relative humidity, temperature, and CO_2. Input your data into Table 30.1.
   B. Determine the pressure (negative or positive) in the room. Input this data into Table 30.1.
   C. Conduct a visual assessment for potential biological contamination. In the room, look for water stains, moisture buildup, and odors that are all potential evidence of water leakage, which can lead to mold. Identify and photograph specific locations and types of potential biological contamination.

3. Select a classroom that was recently occupied. (Be sure that the room has a door that can be closed.)
   A. Inside the recently occupied classroom, test for relative humidity, temperature, and CO_2. Input your data into Table 30.1.
   B. Determine the pressure (negative or positive) in the room. Input this data into Table 30.1.
   C. Conduct a visual assessment for potential biological contamination—water stains, moisture buildup, and odors. Identify and photograph specific locations and types of potential biological contamination.

4. Compare your temperature, humidity, and CO_2 results found in the two classrooms with the American Society of Heating, Refrigeration, and Air Conditioning Engineers (ASHRAE) indoor air quality guidelines. How do your classrooms compare? Are they within the range for the season for each parameter?

5. Examine your building to determine what type of ventilation system is used (e.g., mechanical, windows, combination). Look for any areas where you think there may be an airflow problem. Your instructor might have input on where to go or how to do this, but do not get yourself in any hazardous situations or areas.
   A. Is the building providing adequate ventilation? Why or why not?
   B. What can you say about the infiltration potential and/or air exchanges between the ambient air quality and indoor air quality of the building?

6. Assuming that the two rooms are representative of the building, based on your findings:
   A. What potential sources of indoor air contaminants did you encounter during your inspection?
   B. What recommendations can you make to ensure good air quality in this building considering the need to conserve energy?

Table 30.1 Data Collection Table

Location	Temperature	Relative humidity	CO_2	Biological inspection
Ambient				
Empty classroom				
Occupied classroom				

Table 30.2   Indoor Activity Log

Location	Time spent	Type of activity (and how strenuous)*	Potential contaminants

*Activity is important because as activity increases, the quicker and deeper the breathing is (e.g., sleep vs. exercise) and, thus, increased exposure to potential contaminants.

7. Keep a log to track your own indoor time for 1 day to assess your exposure potential to indoor air contaminants. Using Table 30.2, record the (a) location, (b) amount of time spent, (c) activity, and (d) potential contaminants to which you were exposed. Calculate the percentage of time you spent indoors using the following formula.

$$\left(T_{in(hours)} \div 24 \text{ hrs}\right) \times 100 = \% \text{ Indoors}$$
$$T_{in} = \text{Total time indoors (hrs)}$$

Your lab report will be a technical report. Use the following format:

WRITE-UP

Title:

I. Introduction: Background information explaining indoor air quality. End this section regarding your hypothesized perception of air quality in your building.

II. Methods: Explain who, what, and how you collected your data.

III. Results: Present your findings; be sure to use figures/tables as appropriate.

IV. Discussion: Interpret your results, using your responses to the task questions and be sure to relate your answers to your own experience and to the meaning or applicability of your discussion.

V. References Cited: Be sure to cite references.

U.S. Environmental Protection Agency (US EPA). 1998.
   An Office Building Occupant's Guide to Indoor Air
   Quality (EPA-402-K-97-003). GPO, Washington, DC.

REFERENCES

Lung Power

OBJECTIVES

- Test your lung power as a measure of personal health.

- Compare peak expiratory flow (PEF) in a population sample.

- Formulate a testable hypothesis about human health and PEF.

KEY CONCEPTS AND TERMS

✓ Lung capacity

✓ Peak expiratory flow (PEF)

✓ Spirometry

TIME REQUIRED: 2 to 3 hours.

INTRODUCTION

This lab requires assessing one's own lung peak expiratory flow (PEF) and reporting the results anonymously. However, approval should be obtained from your school's Intuitional Review Board (IRB) prior to the lab in accordance with your institution's policies and procedures.

Asthma was not a common illness until the middle of the 20th century. As of 2011, approximately 39 million Americans had been diagnosed with asthma (American Lung Association, 2012). There is no cure for asthma although there is treatment. Other respiratory illnesses are also increasing due to workplace exposures, home environments, and other factors.

A common measure of "lung power" is **peak expiratory flow**, which is the maximum rate a person can exhale during a short maximal expiratory effort after a full inspiration. PEF can be negatively affected by multiple factors; asthma, smoking, exposure to second-hand smoke, and exposure to air pollutants (e.g., particulates, ozone, silica) can be major influences, as well as other personal health issues and workplace exposures. **Lung capacity** is related to PEF. Normal adult lung capacity is typically from 3 to 6 L and is calculated as milliliters of air per kilogram of body weight (mL/kg_{BW}).[1] Women should have 50 to 60 mL of air per kilogram of body weight (55 mL × kg). Men should have about 70 mL per kilogram of body weight (70 mL × kg) (Bellemare, Jeanneret, and Couture, 2003). Age is also a factor considered in determining lung capacity. **Spirometry** is the measurement of lung function through a pulmonary function test (PFT). We can get an idea about respiration by using a simple device—a PEF meter—the same tool used by millions as part of their personal health-care regime.

MATERIALS

- Excel or similar software

- Internet access for site research

- Peak flow meter (one per student; cannot be shared)

---

[1] 1 mL is equal to 1 cc.

As a class, design (or agree on) a method for sampling respiratory flow rate. For example, you may decide to sample thrice a day for a 2-week period. Decide the environmental and personal factors to be recorded. Your data will be collected by you, and then inputted anonymously. However, to best protect privacy, the class should agree on the nature of the data collected and how it is to be reported. Decide also what graphs will be used to present the data and what statistics will be used to analyze the data.

> **TASKS**

1. Using a peak flow meter (a portable device that measures airflow or the PEF rate), record the results. Then, repeat two more times and record the highest number of all three measurements.

2. As a class, pool everyone's peak flow measurement. Use Excel to graph the data. As a class, discuss the data and what sorts of hypotheses could be tested.

3. Construct a hypothesis that involves a class variable (demographic variables such as height, weight, age, gender, and ethnicity) for comparison with the flow rates. Be sure to cite at least one source from the literature in your rationale for your hypothesis.

4. Write a formal laboratory report based on your findings. Discuss the implications of the results. What recommendations do you have for further research?

> **PEAK FLOW METER USE**
>
> 1. Make sure device reads zero or is at base level.
> 2. Stand up (unless you have a physical disability).
> 3. Take as deep a breath as possible.
> 4. Place meter in your mouth and close your lips around the mouthpiece.
> 5. Exhale as hard and as fast as possible into the meter.
> 6. Do not let your tongue block the mouthpiece.

A formal experiment laboratory write-up is required using the headings below. In writing your discussion section, be sure to address the following questions in a narrative format.

> **WRITE-UP**

Title:

I. Introduction: What is your control? What was your variable? What support do you have for your hypothesis? Be sure to end this section with your hypothesis.

II. Methods: Explain what you did. Include a figure showing your setup.

III. Results: Include a results table.

IV. Discussion:
- Did you accept or reject your hypothesis?
- Relate your findings to environmental health implications. For example, how prevalent is asthma in your state/community/population?
- What do you think your results mean?
- What possible confounding (interfering or overlapping) variables might make one trial different from another (e.g., gender, size, age, stress)?
- Discuss what variables might make the control yield different experimental results from the trials. Discuss to what degree you can control these variables.
- If you were doing it again, discuss what changes you would make in setting up and conducting your experiment.

V. References Cited

REFERENCES

American Lung Association, Epidemiology and Statistics Unit, Research and Program Services. 2012. Trends in Asthma Morbidity and Mortality. New York. Available at http://www.lung.org/assets/documents/research/asthma-trend-report.pdf (verified 15 August 2017).

Bellemare, F., A. Jeanneret, and J. Couture. 2003. Sex Differences in Thoracic Dimensions and Configurations. American Journal of Respiratory Critical Care Medicine. 168(3): 305–312.

Applied Problem Sets

The Scientific Method: Observation and Hypotheses

By now, you have examined a variety of phenomena in the natural world. A focus of environmental science is to explain these phenomena. That is, what are the causes and effects of such phenomena? How does one go about investigating the causes and effects such that the conclusions are meaningful and reliable? The answer is the **scientific method**, which uses designed experiments and careful observations to investigate causes and effects.

The scientific method is the systematic procedure for investigation, which generally is composed of the following:

1. Observe a phenomenon.

2. Formulate a hypothesis (H_0 and H_A)—a tentative description that explains your observation.

3. Design and conduct an experiment to test the hypothesis—the collection of data to test the hypothesis.

4. Present your results.

5. Interpret and discuss the results that accept or reject your hypothesis, and then modify the hypothesis as necessary.

STEP 1— OBSERVE A PHENOMENON

Suppose you observe that an environmental phenomenon (e.g., rainbow trout, *Salmo gairdneri*, are no longer present in a nearby river). You could conduct research by examining the literature to determine what other studies have been done to help you find a possible explanation. However, the studies may not be directly applicable to your particular region. You might want to design and conduct your own study based on your observations, what you know, and what others have researched.

STEP 2— FORMULATE A HYPOTHESIS

The second step in the scientific method is to formulate a hypothesis. A **hypothesis** is a predictive statement of the cause and effect in a specific situation. It is a tentative statement that proposes a possible explanation of an observed phenomenon that is testable.

A well-written hypothesis should:
- Include an independent (manipulated—the cause) and a dependent (resultant—the effect) variable.
- Be testable with a straightforward experiment (*testable* is key because you will need to design and perform an experiment on how two variables might be related).
- Be based on observations or knowledge.
- Be dichotomous—it must be written so that the predicted outcome is either confirmed or not confirmed. It should not contain the words "and," "or," or "also" as these would suggest nondichotomous hypotheses.
- Be unambiguous—it must be clear and specific in its predicted results.

In developing a hypothesis, you must form a set of two contradicting hypotheses: the null hypothesis (symbolized by H_0) and the alternate hypothesis (symbolized by H_A). Basically, the null hypothesis is a stated assumption that there is *no* effect in a cause-and-effect process or relationship. In contrast, the alternate hypothesis is a statement that there is an effect and often predicts what the effect will be. Thus, H_0 is to determine whether a relationship exists and H_A is to determine the direction or nature of the relationship.

Your hypotheses can be as general or as specific as needed. However, a very specific hypothesis is much easier to test in a meaningful way that can yield useful data.

EXAMPLE OF GENERAL HYPOTHESES:

H_0: <u>Dissolved oxygen</u> does not affect **fish**.

H_A: <u>Dissolved oxygen</u> affects **fish**.

As you can see, the hypothesis is very general and is not very meaningful. It is not a well-written hypothesis as defined. What type of fish? What is meant by affect? How much dissolved oxygen will cause an effect? How can it be tested?

EXAMPLE OF SPECIFIC HYPOTHESES:

H_0: Less than 3 mg/L of <u>dissolved oxygen</u> is not lethal to **juvenile Rainbow trout** (*Salmo gairdneri*).

H_A: Less than 3 mg/L of <u>dissolved oxygen</u> is lethal to **juvenile Rainbow trout** (*Salmo gairdneri*).

These hypotheses are testable, unambiguous, dichotomous, and include dependent (bold) and independent (underlined) variables. Also, you know exactly what to test and the results are meaningful in relationship to the hypothesis.

The next step is to design an experiment to test your hypothesis. Your experiment is affected by variables. There are three kinds of variables in an experiment: independent, dependent, and controlled. The **independent variable** is the variable that you purposely manipulate—the cause. The **dependent variable** is the variable that is being observed—the effect, which changes (or might change) in response to the independent variable. Your written hypothesis must contain both an independent and a dependent variable. Variables that are not changed are called **controlled variables**.

> **STEP 3—
> DESIGN
> AND CONDUCT
> AN
> EXPERIMENT**

When conducting an experiment, it must be a controlled experiment. That is, you must compare an "experimental or treatment group" with a "control group." The two groups must be *exactly* the same except for the one variable being tested. For example, in your experiment on dissolved oxygen and trout, you may elect to use two aquariums and five juvenile Rainbow trout. The aquarium, food, light source, location of the aquarium, ambient temperature, water temperature, trout stock, trout gender, age of trout, and so forth all have to be exactly the same. The only difference is that the water in one aquarium has a lower concentration of dissolved oxygen of 3 mg/L and the other has a "normal" concentration. The aquarium with "normal" dissolved oxygen is called the **control group** and the aquarium with the lowered dissolved oxygen is called the **treatment group**. The control group acts as a reference point for comparison with the treatment group. If the test is properly designed and constructed, any difference between the two groups can only be due to the one experimental factor (independent variable) you manipulated. The challenge is to control for variables that may affect (confound) your results. A **confounding variable** is an extraneous (uncontrolled) variable that could produce an alternative explanation for the results. Again, controlling confounding variables is done by ensuring that the two aquarium tests are the same except for the treatment—in this case, the dissolved oxygen.

In conducting your experiment, how can you be certain of the results? Was it by chance that the experiment ended the way it did? The key component is **replication**. Your experiment should be repeated several times to reduce the likelihood that the results are by chance. For example, in your dissolved oxygen experiment, you should use multiple fish in each aquarium *and*, repeat the same experiment at least twice.

EXERCISES

1. **Hypothesis Formulation**
   A. Research the meaning of *null hypothesis*. Describe how and why it is used in experimental design. Properly cite your reference.
   B. The following are observations of environmental phenomena. Rewrite each observation into a formal, testable null hypothesis *and* a formal, testable alternate hypothesis. Remember that independent means what you can control and dependent means what change you think might occur depending on the controlled (independent) variable. Put the dependent variable in **bold** font and <u>underline</u> the independent variable for each.
      - Eggshells produced by birds that were exposed to mercury seem to be thinner.
      - Plants near roads where salt is used in the winter appear smaller.
      - Gulls covered with oil do not look like they can fly.
      - More algae is in the water during the summer.
      - Trees in areas where there is acid precipitation look shorter.
   C. Describe five environmental phenomena affecting your region.
   D. For each of these environmental phenomena, create formal, testable hypotheses (both a null and an alternate hypothesis). **Bold** the dependent variable and <u>underline</u> the independent variable for each.

2. **Experimental Design**
   A. From exercise 1.D, choose one of the phenomena and its hypotheses (H_0 and H_A). Use lines and arrows to outline and connect the steps (this is a "block-flow diagram") in your experimental design to test the hypothesis based on the scientific method. (In other words, graphically depict the steps involved in testing the hypothesis.)
   B. List the method including materials and equipment needed for your experimental design. Provide enough specific details (step by step) to explain how the experiment would be conducted.
   C. Write a critique of your experimental design. Your critique should include (a) likely confounding variables, (b) how or what could be done to improve the design, and

      (c)   how   to   refine   the   hypothesis.

The Scientific Method: Results and Discussion

In *Problem Set 1*, you studied the basics of the scientific method—the systematic procedure for investigation, which is generally composed of the following:

1. Observe a phenomenon.

2. Formulate a hypothesis (H_0 and H_A)—a tentative description that explains your observation.

3. Design and conduct an experiment to test the hypothesis—the collection of data.

4. Present your results.

5. Interpret and discuss the results that accept or reject your hypothesis and then modify the hypothesis.

In *Problem Set 1*, you also conducted exercises to formulate a proper, testable hypothesis and to design an experiment to test your hypothesis. Following the completion of your experiment, you will have generated data. In this problem set, you will focus on steps 4 and 5 of the scientific method: *present your results* and *interpret your results*.

The data obtained from your experiment are presented in the results section. Because the data are generally numbers, they are presented in tables or figures. Note that in the results section, you *do not* interpret the results. This is done in the discussion section. In the results section, you must refer to ("call out") your tables or figures. Tables and figures should not stand alone. For example, you might say, "*The results of the experiment are contained in Table 1*" rather than just "*see Table 1.*"

Now that you have called out (i.e., referenced) Table 1, you must present Table 1 as soon as possible in your paper (generally following the paragraph that contains the call-out). The same applies to figures. Thus, they are not attachments or appendices, but are inserted directly into the report.

The preliminary data collected from your experiment is referred to as **raw data**. Raw data has not been processed (extracted, organized, formatted, summarized, or analyzed) for presentation. The protocol is to summarize and organize data before presenting it in the results section. For example, you may want to present basic, descriptive statistics such as mean or mode, or percentages. Consequently, you will need to perform some basic calculations.

Data are presented in tables as numbers and symbolically in figures (charts and graphs), which are used to depict trends or proportions. Tables and figures can be produced in word processing, spreadsheet, or statistics programs. As presented below, the major data presentation formats are tables, histograms, and line graphs. Note that titles are placed on top of tables and below figures (histograms and line graphs).

TABLES

Tables are used to present data in a tabular or spreadsheet format. See Table P2.1 for an example.

Table P2.1 Total and Average Number of Tropical Cyclones by Month, 1851–2015 (NOAA, 2016)

Month	Hurricanes
January	2
February	0
March	1
April	0
May	4
June	33
July	55
August	241
September	398
October	203
November	59
December	6

BAR CHARTS

A bar, or column, chart depicts categorical frequencies. Generally, the independent variable is plotted along the x (horizontal) axis and the dependent variable is plotted along the y (vertical) axis. The independent variable can attain a finite number of discrete values (e.g., 10) rather than a continuous range of values. For continuous values, line graphs are used. See Figure P2.1 for an example of a histogram.

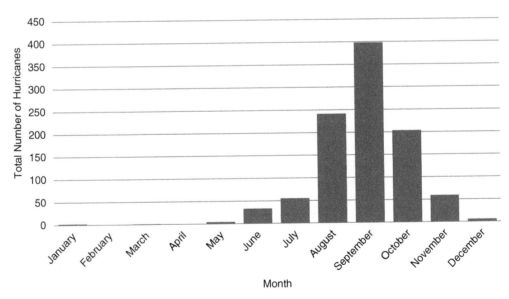

FIGURE P2.1 Total and average number of tropical cyclones by month, 1851–2015

Courtesy: U.S. National Oceanic and Atmospheric Administration (NOAA, 2016).

LINE GRAPHS

A line graph is used to depict the trend of one or more items over a period of time or number of events. Figure P2.2 is an example of a line graph of tropical storm occurrence.

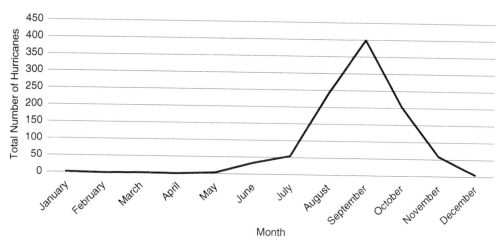

FIGURE P2.2 Total and average number of tropical cyclones by month, 1851–2015

Courtesy: U.S. National Oceanic and Atmospheric Administration (NOAA, 2016).

STEP 5—
INTERPRET
AND
DISCUSS THE
RESULTS

The final step in the scientific method is to interpret and discuss your results. In this step, you examine the results and state whether your null hypothesis is accepted or rejected. If your hypothesis is accepted, briefly explain the major factors and data that support it. If, however, your hypothesis was not accepted, explain your results: Was there a flaw in your experimental design? Were there confounding variables that may have affected your results? Was the data valid, but simply did not support the hypothesis? An important aspect of the scientific method is that rejecting your hypothesis is perfectly acceptable because the purpose of the scientific method is to learn something from the process. If you reject your hypothesis, you can refine your hypothesis or experimental design and then retest.

To support your hypothesis, you must ask whether the results are significant. Ask yourself whether there are tests that you might use to determine if the results are significant.

Notice that the word "prove" was not used. In science, you do not prove anything because science is based on **induction**—a type of reasoning where one uses specific examples or facts to make generalizations or hypotheses. The inductive method entails drawing generalized conclusions based on evidence. Thus, scientists can only draw conclusions on what they find, not on what they cannot find. In other words, there is always the untested hypothesis. This does not mean that scientists lack confidence in the findings, but rather that there is recognition that an alternative explanation may always be possible, even though there may be a very slim chance of this occurring. Instead of using the word *prove*, scientists tend to use words such as the following: supports, agrees with, suggests, affirms, upholds, and demonstrates.

EXERCISE

1. **Presenting Results**
   A. Create a computer-generated table using the following data. Be sure to include a title and labels for the columns, and cite the source and present the full citation in your reference section. The data is the top 10 causes of impairments (degradation of the designated water quality standards) in the United States (US EPA, 2017a): Mercury, 4,511; pathogens, 9,717; pH, 4,409; nutrients, 7,123; metals (other than mercury), 6,843; unknown, 4,170; oxygen depletion, 6,584; temperature, 2,966; polychlorinated biphenyls, 6,049; sediment, 5,969.

B. With the above data, create a bar chart with the computer. Discuss the visual impact of the table compared to the bar chart. Be sure to include a title and labels for the columns, and cite the source and present the full citation in your reference section.

C. Use the computer to create a line graph using the following data, which are the mean, annual ambient concentrations of CO_2 (ppm) measured at Mauna Loa, Hawaii (Tans and Keeling, 2017). Be sure to include a title and labels, and cite the source and present the full citation in your reference section. What does the data suggest is likely to happen regarding future CO_2 levels?

2000	369.55
2001	371.14
2002	373.28
2003	375.80
2004	377.52
2005	379.80
2006	381.90
2007	383.79
2008	385.60
2009	387.43
2010	389.90
2011	391.65
2012	393.85
2013	396.52
2014	398.65
2015	400.83
2016	404.21

D. In accordance with the federal Emergency Planning and Community Right-to-Know Act, specified industries (e.g., companies that manufacture or process 25,000 lb per year or otherwise uses 10,000 lb per year of some 600 designated toxic chemicals) must report releases of these designated toxic chemicals during the previous calendar year. According to the reports submitted to the US EPA (2017b), the following data contains the top 10 industrial releases of designated toxic chemicals in 2016 (in pounds) by state.

Alaska, 833,848,953; Arizona, 85,055,785; Illinois, 109,549,325; Indiana, 129,990,697; Louisiana, 142,811,333, Nevada, 316,820,585; Ohio, 96,491,263, Texas, 201,224,890; and Utah, 271,384,698.

I. Convert this data into kilograms and place the converted data into a bar chart. Be sure to include a title and labels, and cite the source and present the full citation in your reference section.

II. Place the converted data into a pie chart. Again, be sure to include a title and labels, and cite the source and present the full citation in your reference section.

III. Comment on the differences between these two graphic forms of presentation. Does each suggest a different meaning for the data? What is the most appropriate use of each chart?

2. Exercise—Discussion

A. Describe any discernible patterns in the surface water data from question 1.A and in the air data from question 1.D.

B. Your hypothesis was that industry is the leading cause of stream impairment in the United States. Based on the data presented in question 1.A:

  I. Can you support this hypothesis?

  II. What are some potentially confounding variables?

  III. What could you do to further increase your confidence in your results?

C. Your hypothesis was that more populous states release more toxic chemicals than less populous states. Based on the results presented in question 1.D:

  I. Can you support this hypothesis?

  II. What are some potentially confounding variables?

  III. What could you do to further increase your confidence in your hypothesis?

D. Explain why is it important to be able to interpret environmental information using graphs.

REFERENCES

National Oceanic and Atmospheric Administration (NOAA). 2016. Total and Average Number of Tropical Cyclones by Month, 1851–2015. Available from http://www.aoml.noaa.gov/hrd/tcfaq/E17.html (verified 18 October 2017).

Tans, P., and Keeling, R. 2017. Mauna Loa CO_2 Annual Mean Data. National Oceanic and Atmospheric Administration (NOAA). Available from https://www.esrl.noaa.gov/gmd/ccgg/trends/data.html (verified 24 October 2017).

U.S. Environmental Protection Agency (US EPA). 2017a. Causes of Impairment for 303(d) Listed Waters. Available from https://iaspub.epa.gov/waters10/attains_nation_cy.control?p_report_type=T#causes_303d (verified 24 October 2017).

U.S. Environmental Protection Agency (US EPA). 2017b. TRI Explorer. Available from https://iaspub.epa.gov/triexplorer/tri_release.geography / (verified 25 October 2017).

Quantification of Environmental Problems

In this problem set, you will explore the fundamental components of expressing and quantifying environmental problems: scientific notation, calculating percent change over time, metric conversions, and spreadsheet applications.

SCIENTIFIC NOTATION

Environmental science often deals with very large numbers (e.g., the biomass production of a forest) and very small numbers (e.g., the contamination of an aquifer). To better manage these data, and reduce the number of errors, scientists developed a method to express very small and large numbers using *scientific notation*. Scientific notation is based on powers of the base number 10. Thus, for example, an environmental scientist calculates that 146,000,000,000 kg of biomass were produced in a test plot during the previous year. Using scientific notation, the number would be

$$1.46 \times 10^{11}$$

To write a number in scientific notation for 146,000,000,000:

Step 1: Place the decimal after the first digit and drop all of the zeroes.

$$1.46,000,000,000$$

Step 2: Count the number of places from the decimal to the end of the number. For example:

1	.	4	6	0	0	0	0	0	0	0	0	0
Count	.	1	2	3	4	5	6	7	8	9	10	11

There are 11 places after the decimal point; therefore, the exponent is 11. Thus, 146,000,000,000 becomes 1.46×10^{11}.

$152,000,000$ is 1.52×10^8

Exponents in scientific notation are often expressed in different ways. For example, because calculators and spreadsheets generally do not display powers, an "E" is used to represent "$\times 10^x$." Thus, 146,000,000,000 can also be written as

$$1.46E + 11 \text{ or } 1.46 \times 10 \,\widehat{}\, 11$$

MULTIPLYING, DIVIDING, ADDING, AND SUBTRACTING EXPONENTIAL NUMBERS

Exponential numbers are expressed as a digit term between 1 and 10 multiplied by an exponential term. To multiply exponential numbers, first multiply the digit terms, and then add the exponents. For example, 3.1×10^2 multiplied by $4.0 \times 10^8 = (3.1 \times 4.0) \times 10^{(2+8)} = 12.4 \times 10^{10}$. To divide exponential numbers, divide the digit terms, and then subtract the exponents. For example, 3.1×10^2 divided by $4.0 \times 10^8 = (3.1 \div 4.0) \times 10^{(2-8)} = 0.775 \times 10^{-6}$, which we covert into proper exponential notation writing 7.75×10^{-7}. Since we only need two significant figures, we can further simplify our answer to 7.8×10^{-7}.

A simple way to add or subtract exponential numbers is to convert them by moving the decimal place to get the same exponent so that you can just add or subtract the digit terms, then convert back into scientific notation.

A similar approach is used for small numbers (<1), which will have a negative exponent. A millionth of a second is

$$0.000001 \text{ sec.} = \mathbf{1.0 \times 10^{-6}} \text{(or } 1.0E-6 \text{ or } 1 \times 10\text{\textasciicircum}{-6})$$

Thus, with small numbers, you count from the decimal the number of zeroes until you reach the first nonzero number.

$$0.00000123 \text{ sec.} = \mathbf{1.23 \times 10^{-6}}$$

Convert the following numbers into or from scientific notation. Use commas as appropriate for the numbers you convert from scientific notation:

1. 788,950,000,000 _____

2. 1,000,000,000,000 _____

3. 0.00000007832 _____

4. 7,896,000,000 _____

5. 8,718 _____

6. 0.000000045678963 _____

7. 2.223×10^9 _____

8. 3.19E+12 _____

9. $4.444\text{\textasciicircum}{-4}$ _____

10. 7.47×10^{13} _____

EXERCISE SET I: SCIENTIFIC NOTATION

SIGNIFICANT FIGURES

How long should a number be? How many places do I give to the right of the zero? The term *significant figures* refers to the number of digits given in a numerical term. Since conversions and calculations do not necessarily give a number with only a few digits, we use only the digits that are meaningful. For example, a ruler can measure something that is one-third of a meter. But if we divide 1 by 3, we get 0.33333333333333 . . . (we could keep going and list 3s forever). But how meaningful is it to have a million threes? In determining significance, remember that any digit other than zero is automatically significant, and any zero between two significant digits is also significant. Any zero to the right of a decimal point is significant (e.g., 1.010 has four significant figures in it). Exact numbers do not affect accuracy or precision because significant figures are not an issue: an exact number is a defined number or a count of something (e.g., 12 eggs are in a dozen, 3 ft are in a yard, 20 students are in a class; think of the exact number as having no units). When you are working with number conversions and calculations, keep to the same number of significant figures available in the number that has the least amount of significant figures (e.g., 1.001 ÷ 12 = 0.08 rather than = 0.08431667, because the least number of significant figures is in the number 12, which has two significant figures). This will prevent you from being too precise, which implies a greater accuracy than the data or calculations deserve. Spreadsheets and calculators may not be sensitive to this issue, so you may need to round off the resulting number before recording it.

CALCULATING PERCENT CHANGE	Environmental scientists often analyze trends. A common approach to communicating these trends is the percentage of increase/decrease over time. Percentage means "per cent," which means "per hundred." In calculating the percentage of change (whether an increase or a decrease), you are concerned with the difference between two numbers and how much of the first number added to or subtracted from the first number will produce the second number.

The three steps for calculating percent change are as follows:

1. Subtract

2. Divide

3. Multiply

Subtract: New number – Original number = Difference
Divide: By original number
Multiply: By 100

Thus, the basic formula is simple:

$$\frac{\text{Difference}}{\text{Original number}} \times 100 = \%$$

*Percent increase and decrease are calculated with respect to the value **before** the change took place—the original number.*

A. In January, the concentration of acetone in the river was 100 parts per million (ppm). In August, the concentration increased by 12 ppm. What is the percentage increase in acetone?

$$\frac{(112-100)100}{100} = \frac{12 \times 100}{100} = 12\%$$

B. In April, the level of ammonia was 63 ppm. In October, the level was 112 ppm. What is the percentage of increase since April?

Step 1: Determine the amount of change.

$$112 - 63 = 49$$

Step 2: Divide the change by the original number and then multiply by 100.

$$49/63 = 0.777 \times 100 = 77.7\%$$

C. Last month, the level of hexachlorobenzene decreased by 8 ppm. If the level is now 47 ppm, by what percentage did the level of hexachlorobenzene decrease? The amount of change is 8; therefore, the original amount is 47 + 8 = 55.

$$8/55 = 0.145 \times 100 = 14.5\%$$

D. A utility's operating costs for its electrostatic precipitator was as follows:

2017	$345,000
2018	$325,000
2019	$345,000

What were the annual percent changes?

2017 to 2018: [($325,000 – $345,000)/$345,000] × 100 = −5.79%
2018 to 2019: [($345,000 – $325,000)/$325,000] × 100 = 6.15%

Calculate the following:

1. 234.98 to 324.77 = _____% change

2. 324.77 to 234.98 = _____% change

3. 7.14×10^7 to 8.47×10^8 = _____% change

4. 100 increases by 300% = _____

5. 756 declines by 100% = _____

6. 756 increases by 100% = _____

7. A utility's total pollution control costs are broken down as follows:

   Air pollution control = $234,000
   Wastewater treatment = $167,000
   Solid waste = $45,000
   Hazardous waste = $12,000
   What percentage does the company spend for each?
   Air pollution control = _____%
   Wastewater treatment = _____%
   Solid waste = _____%
   Hazardous waste = _____%

8. The company's managers want to budget sufficient money for next year. Assuming that costs will increase by 2.4%, what will be the costs for next year for each category?

   Air pollution control = $ _____
   Wastewater treatment = $ _____
   Solid waste = $ _____
   Hazardous waste = $ _____

9. An environmental scientist has the following data on mercury concentrations in smallmouth bass (*Micropterus dolomieu*) sampled from a pond. What is the annual mean concentration for each year? What is the mean for all 3 years?

2018	2019	2020
7.2×10^4	2.1×10^4	6.4×10^4
6.5×10^4	4.1×10^4	7.7×10^4
7.1×10^4	7.0×10^4	5.2×10^4
6.6×10^4	7.3×10^4	7.2×10^4
6.9×10^4	5.9×10^4	7.1×10^4
7.1×10^4	5.0×10^4	9.4×10^4
7.5×10^4	5.1×10^4	2.1×10^4

10. What is the percentage of difference between 2018 and 2019 and between 2019 and 2020 based on the annual average mercury concentrations?

The universal system of measurement is the metric system. Only three countries have not adopted the metric system, including the United States. However, American scientists use the metric system to communicate their work. Thus, it is important to learn the more logical metric system in environmental science. Below are some basic metric conversions; consult the appendix for the formulas.

EXERCISE SET III METRIC CONVERSIONS

Convert the following:

1. A low-flow toilet averages 2.2 gal per flush. Convert this to liters: _____

2. The 2017 Infinity AWD is rated as getting 20 miles per gallon. Convert this to kilometers per gallon: _____

3. Convert the above kilometers per gallon into kilometers per liter: _____

4. The ecological footprint for the average American is about 25 acres. Convert this to hectares: _____

5. Combusting a gallon of diesel fuel releases 22.2 lb of CO_2. Convert this into grams per liter: _____

6. There are some predictions that sea level rise from global warming will be as high as 35 in. Convert this to millimeters: _____

7. Now convert the above millimeters to centimeters: _____

8. Predicted surface temperature increases from global climate change are estimated to be from 2.0 to 11.5°F. Convert this range to Celsius: _____

9. Research has shown that driving 55 miles per hour saves 15 to 20% fuel. Convert this to kilometers per hour: _____

10. The average American produces 4.4 lb of municipal solid waste per day. Convert this to kilograms: _____

Ecosystem Diagram

Botkin and Keller (2011:83) define **ecosystem** as "two major parts: nonliving and living. The nonliving part is the physical-chemical environment, including the local atmosphere, water, and mineral soil (on land) or other substrate (in water). The living part, called the **ecological community**, is the set of species interacting within the ecosystem." Nutrients cycle through the ecosystem. Sustained life on the earth depends on ecosystems, not on individual species or populations. Sometimes the boundaries of an ecosystem are well defined and sometimes they are vague. An ecosystem may be a large forest or a tiny puddle; whatever the size, it must have the flow of energy and the cycling of chemical elements. An ecosystem can be artificial or natural, or a combination. Drawing an ecosystem is a way to show your understanding of environmental terms and concepts (Sanford, Staples, and Snowman, 2017).

Your task is to provide an illustration that could be used in a high school environmental science textbook. Select a local ecosystem that is particularly familiar to you. Draw the ecosystem. Some things to consider are feedback arrows (positive and negative), energy paths, biogeochemical cycles, habitats, producers, consumers, trophic levels, dominant species, biodiversity, complexity, and organism interrelationships (i.e., predator and prey). This is not an art class, but your drawing does need to be clear and easy to understand and contain a significant amount of factual information.

After you draw the ecosystem, answer the following:

1. Where is this ecosystem found?

2. What roles are filled by this ecosystem in terms of the surrounding environment?

3. What are the principal characteristics of this ecosystem?

4. What are the main forces and mechanisms acting on the species and habitats in this ecosystem?

5. What are the critical resources necessary to sustain this ecosystem?

6. What is the advantage of studying the environment at the *ecosystem* level?

REFERENCES

Botkin, D.B. and E.A. Keller. 2011. Environmental Science: Earth as a Living Planet. 8th ed. Wiley & Sons, New York.

Sanford, R.M., J.K. Staples, and S.A. Snowman. 2017. The Draw-an-Ecosystem Task as an Assessment Tool in Environmental Science Education. *Science Education and Civic Engagement: An International Journal.* SENCER. Winter 2017. Available at http://ncsce.net/the-draw-an-ecosystem-task-as-an-assessment-tool-in-environmental-science-education (verified 7 November 2017).

Biogeochemical Concept Map

Recall the examples of systems and biogeochemical cycles in the your environmental science textbook (or research these concepts on your own). For this homework, you will create a diagram called a "concept map." A concept map is an analytical tool to visually understand complex information; essentially, it is a diagram that visually depicts the various relationships among concepts within a specific topic. The concept map contains nodes or cells, each with a concept, item, or question. The nodes or cells are linked, with the links labeled and direction of flow indicated with an arrow. The labels explain the relationship between the nodes. The arrow reads like a sentence to describe the direction of the relationship.

Create your own concept map for **one** of the following "big six" (macronutrients): carbon, hydrogen, oxygen, nitrogen, phosphorus, or sulfur. Be sure that your concept map addresses time, chemical reactions, pathways, and other appropriate factors. Figure P5.1 is a sample concept map for the hydrologic (water) cycle.

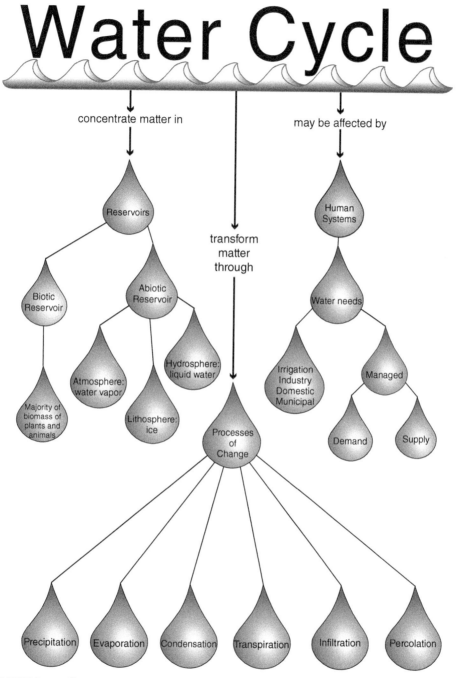

FIGURE P5.1   Sample concept map—the hydrologic (water) cycle

Global Climate Change, CO₂, and You

INTRODUCTION

The economies of the industrialized world are dependent on fossil fuels. Coal, natural gas, and petroleum, formed hundreds of millions of years ago from decayed plants and animals, have provided modern people with a supply of stored energy generated from millions of years of solar energy. Fossil fuels have allowed us to move from a society based primarily on energy from people, living plants, animals, wind, and flowing water to the one based on fossil fuels. Fossil fuels were formed under special conditions that no longer exist and thus are a limited resource.

Limited supplies are, however, not the only concern. When fossil fuels are burned, they produce, among other pollutants, carbon dioxide (CO_2), the principal contributor to the greenhouse effect and global climate change. Anthropogenic activities (fossil fuel combustion and industrial processes) contribute about 78% of the total CO_2 emissions with land-use practices (e.g., agriculture, deforestation) accounting for the second-largest contribution of CO_2 (IPCC, 2014). Globally, in 1970, greenhouse gas (GHG) emissions from fossil fuels were 14.85 billion tons and in 2016, this amount had increased to 31.85 billion tons (IPCC, 2014). The total global emissions of GHGs in 2016 were 49 billion tons. As shown in Figure P6.1, since 1959, the ambient concentration of CO_2 in the atmosphere has increased by about 27.9%. The significance of this substantial increase in ambient CO_2 is the **enhanced greenhouse effect** as depicted in Figure P6.2, which leads to global climate change. Most climate scientists believe that stabilizing the climate will require slashing worldwide CO_2 emissions in half. Because the planet's population was somewhat more than 7.4 billion people, each person's share of GHG emissions in 2016 was 4.3 tons annually.

The purpose of this problem set is to calculate how much CO_2 you contribute through your personal daily activities. To make a rough calculation of your emissions, answer the questions and record the data into Table P6.1. Be sure to keep track of your units.

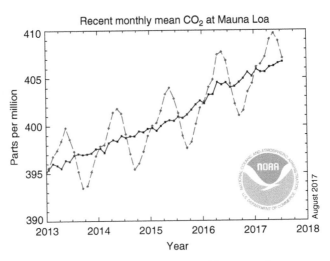

FIGURE P6.1 Monthly mean carbon dioxide measured at Mauna Loa Observatory, Hawaii

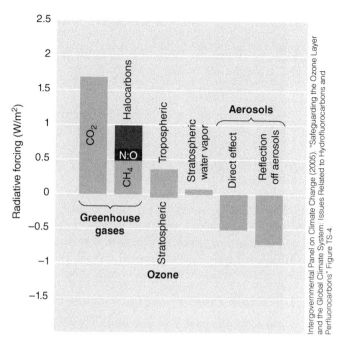

FIGURE P6.2   Enhanced greenhouse effect

The buildup of carbon dioxide (CO_2) and other greenhouse gases warms the atmosphere by absorbing some of the outgoing infrared (heat) radiation. Some of the heat in the warmed atmosphere is transferred back to Earth's surface, warming the land and ocean.

Table P6.1   Personal CO_2 Data

Activity	Annual CO_2 emissions
Motor vehicle	
Air travel	
Electricity	
Heating and cooling	
Subtotal	
Indirect energy consumption	
Total	

1. Estimate the number of miles you drive per year and the average miles per gallon your car gets (if you are not sure, use the national average, which is about 13,500 miles/year).

    _____ miles/year divided by _____ miles/gal = _____ gal/year

    Burning a gallon of gasoline produces 9 kg of CO_2, so multiply the above total gallons per year by 9 to get your total CO_2 emissions from driving.

    _____ gal/year multiplied by 9 kg/gal = _____ kg/year
    Total car CO_2 attributable to you: _____

2. Estimate the number of miles you flew last year. Airplane fuel efficiency varies, but, on average, flying 1 miles generates 0.23 kg (0.5 lb of CO_2) per passenger.

**TRANSPORTA-
TION ACTIVITIES**

Number of miles flown _____ miles/year multiplied by 0.23 kg/mile = _____ kg/year

Assume 100 people on an average flight and divide by that number: _____ kg/year

Total air travel CO$_2$ attributable to you: _____

ELECTRICITY

Estimate the number of kilowatt-hours of electricity you consumed per year. Multiply the number on your last monthly electric bill by 12 to get an estimate for the year. If you do not have an electric bill, go to "estimating appliance and home electronic energy use" (https://www.energy .gov/energysaver/estimating-appliance-and-home-electronic-energy-use) and add the values for each appliance to find your annual energy consumption for electrical appliances in kilowatt-hours per year. While the amount of CO$_2$ generated per kilowatt-hour of electricity varies depending on the fuel used in your state or region. For example, nuclear energy, wind, solar power, and hydroelectric power are effectively free of CO$_2$, but they produce only about 15% of the nation's electricity while fossil fuels (coal and natural gas) are used to produce about 64% of our electricity (EIA, 2017). Therefore, use 0.62 kg per kilowatt-hour as the default.

_____ kilowatt-hours/year multiplied by 0.62 = _____ kg/year

If you have more than one person in your household, divide accordingly to get your personal share from the total: _____ kg/year ÷ number in household = _____ kg/year total electricity CO$_2$ attributable to you

(For example, if there are four in your household, assume 25% of the total is your contribution.)

3. Heating and Cooling
   A. If you use natural gas as an energy source, estimate the number of BTUs of natural gas you used last year. Multiply the number on your last monthly natural gas bill by 12 (months/year) to get an estimate for the year. If you do not have a natural gas bill, assume a typical value of 60,000,000 BTUs/year. 100,000 BTUs of natural gas produce 5.5 kg of CO$_2$. 100,000 BTUs is approximately the yield of 100 ft^3 of natural gas. 100 ft^3 of natural gas = 1 ccf (also called 1 Therm).

   Using your natural gas bill or the typical value, determine how many kilograms of CO$_2$ are emitted per year: _____

   What percentage of this is yours? _____

   B. If you use heating oil as an energy source, estimate the number of gallons of oil you used last year. If you do not have an oil bill, assume a typical value of 600 gal/year. 1 gal of oil generates 7.6 kg CO$_2$.

   _____ gal/year multiplied by 7.6 kg CO$_2$/gal = _____ kg/year

   Divide by the number of people in your household: _____ kg/year

   C. If your home is heated by wood, one cord of wood emits 3,068 kg of CO$_2$.

   Add the totals of the above calculations:

   Total heating and cooling CO$_2$ attributable to you: _____

INDIRECT ENERGY CONSUMPTION

4. The amount you calculated above is the energy you use directly. It does not include the energy you use indirectly. Indirect energy use includes energy used in manufacturing the products you buy, growing and processing food, and transporting food and products to retailers. Approximately 75% of the energy we use is used indirectly. Therefore, you need

to multiply the total kilograms per year that you calculated for direct energy consumption by 4 and add this to your direct energy total to obtain your total energy.

_____ kg/year (direct) multiplied by 4 = _____ kg/year (direct and indirect)

Now convert to metric tons.

_____ kg/year total divided by 1,000 kg/metric ton = _____ metric ton/year

5. How does your total compare to the approximate average production of CO_2 per person in the United States of 21 metric tons?

6. For *each* of the above activities (transportation, electricity, and heat and cooling), using the Internet, identify *two* specific actions that can reduce CO_2 emissions through energy efficiency and/or energy conservation.

REDUCING YOUR CO₂ EMISSIONS

Intergovernmental Panel on Climate Change (IPCC). 2014. Climate Change 2014: Synthesis Report, Summary for Policymakers. Available from https://www.ipcc.ch/pdf/assessment-report/ar5/wg3/ipcc_wg3_ar5_summary-for-policymakers.pdf (verified 2 November 2017).

U.S. Energy Information Administration. 2017. Frequently Asked Questions. Available from https://www.eia.gov/tools/faqs/faq.php?id=427&t=3 (verified 2 November 2017).

REFERENCES

Recognizing Human Impacts

This problem set explores relationships among increasing human population, demands for natural resources, and environmental degradation. When you do your calculations, be sure to write the correct units and use appropriate significant figures (see "Significant Figures" textbox in Problem Set 3). Review your answers to be sure they make sense.

SIMPLE MODEL OF ENVIRONMENTAL DEGRADATION

1. A simple model of environmental degradation can be formulated as $ED = P \times A \times T$, where **ED** is environmental degradation, **P** is population size, **A** (affluence) is per capita resource use, and **T** (technology) is environmental degradation per unit resource use.[1] Environmental degradation, ED, increases or decreases with an increase or decrease in any of the three variables (P, A, and T). For a given level of P, unacceptable levels of environmental degradation can result by multiplying very large values of A by small values of T or by multiplying relatively small values of A by very large values of T. Excessive environmental degradation resulting from large values of P has been termed **people overpopulation**, whereas similarly excessive degradation because of large values of A has been termed **overconsumption**. Let's examine some examples of how population size, resource use, and environmental degradation are interrelated.

   A. A substantial portion of the U.S. gross domestic product (GDP) is devoted to the motor vehicle economy (manufacturing, fuel, insurance, repair, road construction, parts, labor, and so forth). Using the Federal Highway Administration's website (2017) (https://www.fhwa.dot.gov), search for state motor vehicle registrations. What is the total number of automobiles and noncommercial trucks registered in the United States?

   B. What is the current population of the United States? Check the U.S. Census Population Clock (https://www.census.gov/popclock) for the current U.S. population.

   C. In 2016, the United States consumed 143.37 *billion* gal of gasoline as motor fuel (EIA, 2017). The combustion of 1 gal of gasoline produces 8.750 kg of CO_2.
      I. What is the number of automobiles per capita for the United States?
      II. What are the total annual CO_2 emissions for U.S. gasoline motor fuel consumption?
      III. What is the per capita generation of automobile CO_2 emissions for the United States (per capita = resource/population)?

   D. What is the current population of the world? Check the U.S. Census Population Clock (https://www.census.gov/popclock) for the current world population. Give the date and time you recorded the current population.

   E. Assuming everyone in the world produced the same amount of CO_2 emissions as the United States does from automotive gasoline, what would the estimated total worldwide CO_2 emissions be?

---

[1] This model is a slight variation of the environmental impact model developed by Ehrlich and Holdren (1971).

2. Measuring environmental degradation ultimately comes down to evaluating how much the quality of life on our planet has declined. In the calculation you made above, you performed relatively straightforward calculations of CO_2 emissions, not direct measures of environmental degradation. Translating CO_2 emissions into environmental degradation is difficult. First, one would have to know the effect of the emissions in terms of increased global climate change and all of its ramifications. Then all of the effects would need to be evaluated and assigned a cost, such as economic cost (cost in dollars) or health cost (loss of life or pain and suffering). Costs such as loss of a species or aesthetic degradation are difficult to quantify. A final evaluation of environmental degradation would ideally tell us how much the quality of life has declined. In a paragraph, describe your idea for the best way to measure environmental degradation. What factors would you consider?

> **QUESTION OF QUALITY**

3. One approach to reducing environmental degradation is to reduce per capita resource use through greater efficiency, energy and/or resource conservation, greater durability of products, reuse, and recycling. The three Rs of sustainable resource use (a.k.a. pollution prevention) can be thought of as reducing unnecessary consumption, reusing when possible, and recycling what you cannot reuse. (Recycling is important, but it is the least environmentally beneficial compared to reduction and reuse.)

> **REDUCING PER CAPITA RESOURCE USE**

  A. A typical older toilet uses 18 to 26 L (5 to 7 gal) per flush (assume 22 L per flush). Modern low-flush toilets use 6 L (1.5 gal) per flush. Assume that each student at a 10,000-student university flushes five times per day.

    I. How many liters of water would be saved in 1 day if all older toilets were converted to low-flush toilets?

    II. How many liters could be saved in a school year (250 days)?

    III. During the 1980s, when a prolonged drought hit California, the following saying became common: "If it's yellow let it mellow, if it's brown flush it down." How do you feel about this?

  B. The average student generates an estimated 290 kg (640 lb) of solid waste (garbage) per year.

    I. Multiply this average amount by the number of students in your campus to determine the total student amount.

    II. What could be done to reduce this amount (recycling addresses postgeneration of waste, the question is how to reduce generation, not disposal)?

4. Population growth (or decline) results from a balance of births, deaths, immigration, and emigration. Let's look at some basic calculations of population growth that do not require detailed information about age, survivorship, or fertility. We will use the exponential growth equation **$dN/dt = rN$**, where **dN/dt** is the rate of population growth, **r** is the intrinsic rate of growth (per capita rate of increase, or birth rate minus death rate) expressed as the number of new individuals per existing individual per time unit, and **N** is the number of individuals in the population. This equation can be solved, using calculus, to obtain the population size as it changes over time:

> **THE POPULATION FACTOR**

  $N_t = N_0 e^{rt}$, where N_t is the population at time t, N_0 is the initial population, e is the base of the natural logarithm (a constant with a value of 2.718281828...), r is the intrinsic rate of growth, and t is the time. For example, Tiny Town has a population of 1,100 and an r of 1.1%. What will the population be in 20 years? $1{,}100 \times (2.718281828^{[(.011)(20)]}) = 1{,}370$. What will the population be in 40 years? $1{,}100 \times (2.718281828^{[(.011)(40)]}) = 1{,}708$.

A. Using the current population of the United States (use your answer from 1.B) and an annual intrinsic rate of increase of 0.81% (r = 0.0081), what is the expected population in 5 years? In 10 years? In 25 years?

B. Assume that there is a rise in the annual intrinsic rate of increase to 0.98% (r = 0.0098). What is the expected population in 5 years? In 10 years? In 25 years?

INDIVIDUALS MATTER

5. It is easy to feel small and helpless—that environmental problems are so big that it does not matter what we as individuals do. At one extreme, this argument could be an excuse not to take responsibility for our actions and to justify doing whatever we want regardless of the environmental impact. But we can make a difference if collectively we act responsibly. As we have seen, small numbers multiplied by large numbers can become very large numbers. The slogan "think globally, act locally" is especially pertinent when working to solve environmental problems. Globally, we use a lot of resources, but per capita resource use is different among different cultures and countries.

A. What would happen if each person on Earth reduced his or her resource use by 25%?

B. Discuss the difference between a 25% reduction for an American compared to a 25% reduction for a Haitian?

C. Would an overall reduction of 25% mean anything in the greater scheme of the world if the current exponential growth of the human population were not controlled? Support your answer.

REFERENCES

Ehrlich, P.R., and J.P. Holdren. 1971. Impact of Population Growth. Science 171: 1212–1217.

Federal Highway Administration (FHWA). 2017. State Motor-Vehicle Registrations – 2015. Available at https://www.fhwa.dot.gov/policyinformation/statistics/2015/pdf/mv1.pdf (verified 2 November 2017).

U.S. Energy Information Administration (EIA). 2017. Frequently Asked Questions. Available at https://www.eia.gov/tools/faqs/faq.php?id=23&t=10 (verified 8 November 2017).

Carbon Footprints and Sustainability

The purpose of this problem set is for you to calculate your own **carbon footprint**, which is defined as the total amount of greenhouse gases produced to directly and indirectly support human activities.

Like all species, humans need certain resources to survive. Humans, however, consume resources not only for survival but also for comfort, luxury, and prestige. Every human-related activity has the potential to contribute to climate change through direct and indirect emissions of greenhouse gases (GHGs), often measured as CO_2 equivalents (CO2e). Direct emissions include, for example, GHGs emitted from the tailpipe of a car. Indirect emissions include the emissions generated from the manufacture of a car; exploring, pumping, transporting, and refining of the petroleum; and the maintenance and subsequent disposal of a car. How can we measure the contribution of a person, organization, city, or country to climate change? One popular measurement is the carbon footprint, which calculates the direct and indirect GHGs generated from a variety of activities.

In this problem set, you will individually analyze your consumption and life activity patterns to calculate your carbon footprint using an Internet-based carbon footprint calculator to answer the following questions:

1. Go to the CarbonFootprint.Com's Household Carbon Footprint Calculator[1] or a site provided or your instructor. Answer all questions on the calculator truthfully. You will need household specific information such as utility bills. If you are not sure of the answer, use the Internet to search for average consumption/values in your region or make an educated guess. For this first part, do not select any measures to reduce your emissions.

   A. What are your annual CO_2 emissions from home energy? How does this compare to the same size household in your region?

   B. What are your emissions from transportation? How does this compare to the national average (using percent difference)? What are your emissions from secondary sources? How does this compare to the national average (using percent difference)?

2. What are your total annual CO_2 emissions? How does this compare to the national average (using percent difference)? The next step is to identify all reasonable and likely measures you could undertake to reduce your CO_2 emissions.

   A. Go back to the House Energy section and identify all reasonable and likely reduction measures. What is the new total after your planned actions? How does this new total compare to the national average (using percent difference)?

   B. Go back to the Transportation section and identify all reasonable and likely reduction measures. What is the new total after your planned actions? How does this new total compare to the national average (using percent difference)?

---

[1] https://www.carbonfootprint.com/calculator.aspx

    C. Go back to the Secondary section and identify all reasonable and likely reduction measures. What is the new total after your planned actions? How does this new total compare to the national average (using percent difference)?

    D. What are your revised total annual CO_2 emissions? How does this compare to the national average (using percent difference)?

3. Go to the U.S. Census Bureau's population calculator at https://www.census.gov/popclock.

    A. What is the current world population? (Note the precise *local* time that you determined this.) Check again exactly 1 hour later (note the precise time that you checked); what is the current world population? How many people were added over the last __ minutes?

    B. Using the most recent world population figure, multiply this by your initial total annual CO_2 emissions and convert into metric tons. Assuming everyone in the world had the same GHG emissions as you, what would the total annual global GHG emissions be?

    C. The total global emissions of GHGs in 2016 were 49 billion tons. How does your answer from 3.B compare to this amount?

    D. What conclusions can you draw from the comparison?

4. Evaluate the carbon footprint calculator you used for this assignment. Discuss how accurate the footprint calculator is for your lifestyle, diet, and transportation. What could be done to improve the accuracy of the carbon footprint calculator?

Oil Consumption and Future Availability

Oil is a nonrenewable resource. In 1859, the first oil well—the Drake Oil Well—was drilled in Titusville, Pennsylvania, initiating the petroleum revolution. On that day, we began to deplete the resource. Today, oil has is a well-entrenched part of the global market. For example, according to the U.S. Energy Information Administration (2017a), in 2016, the United States imported 7.9 million barrels (bbl) of crude oil per day (million bbl/day), produced 14.6 million bbl, and exported 5.2 million bbl/day of crude oil. A barrel is 42 gal and 1 million bbl/day equals 15.33 million gal per year. (Canada is currently the largest exporter of crude oil to the United States and is tied with Mexico for the largest importer of U.S.-produced crude oil.) There are some arguments in the United States that oil is abundant in the ground, but is it usable? If so, who should use it? The current National Energy Policy calls for more exploration and production of petroleum within the United States. The rationale is that it will reduce prices, reduce our dependence on foreign sources of oil, and improve our national security. The questions we must ask: "Is this good environmental policy?" This problem set provides some background before you answer this question with confidence. Answer the questions presented below; be sure to give the source for any data or facts that you cite.

There is a widespread lack of understanding of the significant difference between the terms "resources" and "reserves."

AVAILABLE OIL

The total *resource* base of oil is the entire volume formed and trapped in-place within the Earth before any production. The largest portion of this total resource base is not recoverable by current or foreseeable technology. Most of the nonrecoverable volume occurs at very low concentrations throughout the Earth's crust and cannot be extracted short of mining the rock or by some other approach that would consume *more* energy than it produced.

Reserves, a subset of the total resource base that is of societal and economic interest, are technically recoverable portions of the total resource base. Therefore, reserves are not a fixed amount and are estimated; they can increase or decrease based on geological data, technology, and prices and costs.

A hotly debated subject is whether the United States should open up the Arctic National Wildlife Refuge (ANWR) specifically for petroleum production. The purpose of this problem set is to investigate important facts of the ANWR debate to help you make a more informed decision about the issue of opening up a protected land for energy needs and to answer the question as to whether drilling in ANWR is good environmental policy.

According to the U.S. Energy Information Administration (2008), it will take 10 years from the beginning of the project before oil starts to flow from ANWR. The mean estimated oil reserve of ANWR is 10.42 billion barrels. Mean estimated production in ANWR would reach 780,000 barrels per day in 19 years and then decline to 710,000 barrels per day for the remaining 3 years until depletion. In 2018, the United States consumes 20.34 million barrels of petroleum each day; of this amount, about 46% (9.36 million bbl/day) is gasoline (EIA, 2017b).

1. Assuming that the estimates are correct, and oil from ANWR achieves the 710,000 barrel level 22 years from now, based on current crude oil consumption, what percentage of our total national oil consumption could be supplied by ANWR?

2. Search the Internet and determine the total number of highway vehicle miles traveled in the United States for the most recent year reported. *Hint*: Search for the U.S. Department of Transportation's Bureau of Transportation Statistics.

3. Search the Internet and determine the average fuel efficiency (MPG) of motor vehicles in the United States for the most recent year reported. *Hint*: Search for the U.S. Department of Transportation's Bureau of Transportation Statistics.

4. Create a table listing the total number of highway vehicles traveled every year for the past 20 years to the most currently reported year and determine the overall percent change. In this same table, list the annual MPG for cars and trucks combined over the past 20 years and determine the overall percent change.

5. A gallon of crude oil does not mean a gallon of gasoline. The conversion of oil to gasoline through distillation at a refinery depends on several factors, including the quality of the crude oil, the efficiency of the refinery, the product mix desired by the refiner (e.g., diesel fuel, home heating oil, jet fuel), and so forth. In general, in a 42-gal barrel of crude oil, about 46% (about 19.5 gal) of the oil is refined into gasoline, with the rest refined into diesel fuel, jet fuel, and other petroleum products (EIA, 2017a). Based on this percent and all the information in this problem set, determine how many years ANWR would be able to supply our gasoline (*not* total petroleum) needs if it were our sole source. Support your answer with calculations and clearly identify any assumptions that you made.

6. Based on your answer to Question 5, discuss drilling in ANWR purely from an energy supply perspective.

7. Find a nongovernmental website that *supports* drilling in ANWR. What specifically are the arguments in support of drilling? (Be sure to provide the name of the organization and the URL.)

8. Find a nongovernmental website that *opposes* drilling in ANWR. What specifically are the arguments against drilling? (Be sure to provide the name of the organization and the URL.)

9. Based on your answers, support a position (pro or con) regarding drilling in ANWR from an energy *and* an environmental perspective. To answer the question, is drilling in ANWR good environmental policy?

REFERENCES

U.S. Energy Information Administration (EIA). 2008. Analysis of Crude Oil Production in the Arctic National Wildlife Refuge. Available at https://www.eia.gov/analysis/requests/2008/anwr/pdf/sroiaf(2008)03.pdf (verified 8 November 2017).

U.S. Energy Information Administration (EIA). 2017a. Oil, Crude and Petroleum Products Explained. Available at https://www.eia.gov/energyexplained/index.cfm?page=oil_imports (verified 2 November 2017).

U.S. Energy Information Administration (EIA). 2017b. U.S. Petroleum and Other Liquids Supply, Consumption, and Inventories. Available at https://www.eia.gov/outlooks/steo/tables/pdf/4atab.pdf (verified 2 November 2017).

Water Quality and Consumer Choice

In considering environmental impacts, we can use the precautionary principle as a basis for action. The **precautionary principle** generally states that when an activity has the potential to cause significant threats of harm to the environment and/or human health, we should act even if we have not yet fully identified or understood the cause-and-effect relationships.

This problem set is designed to introduce you to emerging water quality issues and the role of the precautionary principle in the decisions we make about what products to produce and/or consume.

A primary goal of environmental science is to identify and examine emerging environmental concerns before they become critical ecological or human health problems. In 1996, the book *Our Stolen Future* (Colborn et al., 1996) raised concerns about the prevalence of certain synthetic chemicals in the environment that can interfere with hormonal messages involved in the control of growth and development in humans and nonhumans, especially in the fetus. (These are referred to as endocrine disruptors or hormone mimickers.) This book helped focus attention on these and other ubiquitous pollutants in the environment, which had not been studied to the same degree as the so-called conventional water pollutants because of their expected low levels. These overlooked environmental contaminants include reproductive hormones, steroids, antibiotics, pharmaceuticals, personal care products, detergents, disinfectants, fragrances, insect repellants, and fire retardants. These compounds and their metabolites have been detected in U.S. and European waters. (**Metabolites** are byproducts produced from metabolism—the organic processes necessary to sustain life.)

> ### INTRODUCTION

Although environmental scientists have been aware of the problem, until spring 2008, little attention had been paid in the media regarding the potential impact of low levels of pharmaceuticals and personal care products in the environment on ecological or human health. However, a series of stories in the media revealed that low concentrations of endocrine disruptors were present in the surface waters of the United States. These compounds enter the environment directly through discharges from public sewage treatment plants and indirectly through wet weather runoff from animal feed lots and excreta from medicated domestic pets (Daughton, 2003).

According to Kolpin et al. (2002, p. 1202), "Surprisingly, little is known about the extent of environmental occurrence, transport, and ultimate fate of many synthetic organic chemicals after their intended use, particularly hormonally active chemicals, personal care products, and pharmaceuticals that are designed to stimulate a physiological response in humans, plants, and animals." A contributing factor to this lack of data is the limited ability of analytical methods capable of detecting these compounds at the extremely low concentrations expected in the environment. The environmental presence of these compounds raises many concerns (e.g., abnormal physiological processes, reproductive impairment, cancer, and toxicity), but perhaps the biggest concern is that some of these compounds can contribute to the development of antibiotic-resistant bacteria. If super bacteria evolve through resistance to antibiotics, how would we defend ourselves from harmful bacteria?

Kolpin et al. (2002) conducted the first nationwide reconnaissance of the occurrence of pharmaceuticals, hormones, and other organic wastewater contaminants in water resources.

Using a list of 95 contaminants, they identified 139 streams susceptible to contamination across 30 states during 1999 and 2000. A more recent analysis found that 80% of the assessed streams contained one or more of these contaminants. Some of the most frequently detected compounds are as follows:

- **N,N-diethyltoluamide** (also known as DEET, used as a topical insect repellant)

- **Caffeine**

- **Tris (2-chloroethyl) phosphate** (fire retardant used in plastics)

- **Triclosan** (antimicrobial disinfectant used in some soaps, toothpastes, and mouthwashes)

- **4-Nonylphenol** (an ingredient in laundry, dish, and other detergents)

However, these substances are not necessarily endocrine disruptors. Federal agencies such as the US EPA currently are developing tests to use for classifying chemicals as endocrine disruptors.

Some known endocrine disruptors include the following:

- **Bisphenol A** (chemical produced in large quantities for use in polycarbonate plastics and epoxy resins)

- **Di(2-ethylhexyl)phthalate, or DEHP** (a high-production-volume chemical used in the manufacture of a wide variety of plastic-based consumer products and food packaging, some children's products, and some polyvinyl chloride medical devices)

These chemicals can enter the body directly through the use of products (e.g., DEET, caffeine, triclosan, and 4-nonylphenol) or indirectly through the packaging used to contain products (e.g., Bisphenol A, DEHP). Moreover, most of the chemicals can be excreted from the body and enter waterways where they could be consumed again. We lack the information to state with confidence how serious this problem is and to what extent it is causing human or environmental harm. However, we continue to manufacture, consume, and release these products as if there is little or no potential impact.

A crucial part in this process is the consumer. For this assignment, we will focus on identifying chemicals of concern in consumer items and assess the degree to which typical consumers can make an informed decision regarding a particular product and its potential impact on their health, their neighbor's health, and/or the environment.

QUESTIONS

1. Go to a local pharmacy or grocery store. Select a representative product for each of the seven types of consumer products presented in Table P10.1 and complete the table. (Ideally, select an item that you may normally buy based on brand preference and/or price.)

2. Using the product's label, how much information is provided on the ingredients? That is, can you tell what the ingredients are or their potential environmental or health risk? Is the information sufficient to make an informed decision (regarding environmental and personal safety) when purchasing the product?

3. On average, how much (e.g., in grams or liters) of each of the seven *types* of products (include any and all brands of the product that you use) in the above table do you use per year?

4. How much is used in the United States? Multiply these amounts by the current population of the United States for a rough approximation of the total U.S. consumption of these products per year.[1] Comment on the validity of this technique for national estimation.

---

[1] Check the U.S. Census Population Clock (https://www.census.gov/popclock) for the current U.S. population.

Table P10.1 Common Consumer Products

Product	Brand	List the first five ingredients	Identify all health and/or environmental warnings listed on the item
Room deodorizer			
Oven cleaner			
Liquid hand soap			
Skin lotion			
Insect repellant			
Mouthwash			
Toothpaste			

5. For *each* of these seven product types, describe (or construct a block flow diagram) depicting their hydrologic cycle—the process of how they enter the surface water (step by step) during and after use.

6. Select three ingredients of the products that intrigue you and search for information about them on the Internet. Explain what they are and summarize their health or environmental effects. Cite your references

7. Based on your answers, comment on the ability of a typical consumer to make an informed environmental/public health choice using the product's labels.

8. Recommend an alternative to increase the environmental and health information on a label that would not require excessive space, yet would be useful for the consumer.

9. Look up the term "precautionary principle." What does it mean? How would you apply the concept of the precautionary principle to the unrestricted allowance of endocrine disrupters in consumer products? Cite your references

REFERENCES

Colborn, T., D. Dumanoski, and J.P. Myers. 1996. Our Stolen Future: Are We Threatening Our Fertility, Intelligence, and Survival? Penguin Books, New York.

Daughton, C.G. 2003. Environmental Stewardship of Pharmaceuticals: The Green Pharmacy. In Proceedings of the 3rd International Conference on Pharmaceuticals and Endocrine Disrupting Chemicals in Water, National Ground Water Association, 19–21 March 2003, Minneapolis, MN.

Kolpin, D.W., E.T. Furlong, M.T. Meyer, E.M. Thurman, S.D. Zaugg, 2002. Pharmaceuticals, Hormones, and Other Organic Wastewater Contaminants in U.S. Streams, 1999–2000: A National Reconnaissance. Environmental Science and Technology. 36(6): 1202–1211.

Local Environmental Risk

Environmental risk is the ability of an agent (chemical, physical, or radioactive) to cause harm to human health and/or the environment. This problem set is designed to introduce you to the possible environmental risks present in your own community.

Go to U.S. EPA's Envirofacts Data Warehouse: http://www.epa.gov/enviro and do the following:

- Select the tab, **Multisystem Search.**
- Under the **Geography Search**, input a zip code or city name.
- Create a table similar to the following:

Source	Present in Zip Code (Y/N)	Number of Facilities	Distance (meters) to Nearest Receptor	Receptor Types	Physical or Geographical Barriers to Exposure	Physical or Geographical Features Increasing Exposure
Air						
Toxics						
Waste						
Water						

Answer the following questions and place the answers in the table. This will form the basis for your risk assessment report in Question 7.

1. How many local facilities produce and release air pollutants?

2. How many local facilities have reported releases of toxic chemicals?

3. How many local facilities have reported hazardous waste activities?
   A. How many of these facilities are classified as Large Quantity Generators?
   B. How many of these facilities are classified as Small Quantity Generators?
   C. How many of these facilities are classified as Transporters?
   D. How many sites are classified as potential hazardous waste sites designated under Superfund? What types of facilities were they?

4. How many facilities are regulated by U.S. Environmental Protection Agency (US EPA) regulations for radiation and radioactivity?

5. What is the name of the watershed in the area you selected for the above questions?
   A. How many facilities with permits to discharge to waters of the United States?
   B. How many Community Water Systems are there that serve the same people year-round (e.g., in homes or businesses)?

Risk mapping

6. Using the map, identify the location of a sensitive population such as a school and an ecologically sensitive area such as wetland, park, or protected habitat. What is the name of the school? What is the name of the ecologically sensitive area?

   A. What is the distance in meters from your school to the nearest air, toxics, waste, radio-active, *and* water facility?

   B. What is the distance in meters from the nearest air, toxics, waste, radioactive, *and* water facilities to your sensitive ecological area?

Risk assessment report

  7. In one page, assess the environmental and health risk in the zip code or city associated with the facilities you identified using the Envirofacts Data Warehouse. Include the types of facilities, the types of potential releases, and your school and ecologically sensitive area.

  8. What do you think should be done as part of managing or responding to these risks?

Society and Waste

As you will see below, the average American produces a substantial amount of waste commonly referred to as municipal solid waste (Figure P12.1). However, this is not the only category of waste we produce; there is also a far greater amount of industrial waste (hazardous and nonhazardous) created each year. This problem set is designed to introduce you to the amounts and types of waste generated in the United States and in your own community.

**QUANTIFYING
THE GENERATION
OF WASTE**

1. Relate population to municipal solid waste by answering the following.

   A. What is the current population of the United States? Check the U.S. Census Population Clock (https://www.census.gov/popclock) for the current U.S. population.

   B. In 2014, U.S. residents, businesses, and institutions produced 258×10^6 tons of municipal solid waste (MSW) (US EPA, 2016). How many pounds are generated per person (per capita) per day? Per year?

   C. In 2014, the percentage composition of various categories of MSW by weight was (US EPA, 2016):
   Paper: 26.6%
   Yard trimmings: 13.3%
   Food scraps: 14.9%
   Plastics: 12.9%
   Metals: 9.0%
   Rubber, leather, and textiles: 9.5%
   Wood: 6.2%
   Glass: 4.4%
   Other: 3.2%

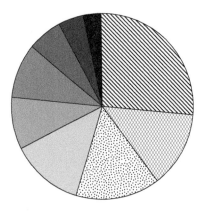

☒ Paper ☒ Yard trimmings ☒ Food scraps
☐ Plastics ☐ Metals ☐ Rubber, leather, and textiles
☐ Wood ☐ Glass ☐ Other

FIGURE P12.1 Composition of municipal solid waste

Source: US EPA, 2016.

D. Using the information in A and B, determine the number of tons of each category of MSW produced in 2014. Then, construct a table with the headings as shown below. Complete the table by giving categories, amounts, and three specific examples of waste in each category.

Category	Amount (in tons)	Examples

2. In the United States, industry generates about 33.646 million tons of hazardous waste (corrosive, reactive, toxic, or ignitable waste) each year (US EPA, 2017).

HAZARDOUS WASTE

   A. Based on the current U.S. population, what is the per capita generation rate in pounds per day? Per year?

   B. How does this compare (quantitatively) to MSW (i.e., percent difference)?

3. Each year, U.S. industry generates an estimated 7.6 billion tons of nonhazardous industrial waste. Most of this waste is disposed of in landfills. Generated by a broad spectrum of U.S. industries, industrial waste is process waste associated with manufacturing and is not classified as either municipal waste or hazardous waste by federal or state law.

NONHAZARDOUS INDUSTRIAL WASTE

   A. What was the per capita generation rate of nonhazardous industrial waste in pounds per day? Pounds per day?

   B. How much is generated *per person* in one day for *all* waste combined (MSW, hazardous, and nonhazardous industrial waste) in (a) pounds per day and (b) tons per day?

4. Pollution prevention is a hierarchical approach to waste management based on the three Rs—Reduce, Reuse, and Recycle. For example:

POLLUTION PREVENTION

   Reduce = Double-sided copying paper
   Reuse = Using paper for scrap paper
   Recycle = Collect used paper to recycle into new paper

   In a table, for *each* waste category in 1.D, identify an action to reduce, reuse, and recycle.

Reduce, or source reduction, means to reduce or eliminate the creation of pollutants (i.e., reduce the volume and/or toxicity) through equipment or technology modifications, process or procedure modifications, reformulation or redesign of products, substitution of raw materials (with less hazardous materials), and improvements in housekeeping, maintenance, training, or inventory control.

Reuse means the reuse of a material without processing the material. This does not mean that the material has to be reused for its original intent; it can be reused for other purposes provided that it does not undergo significant processing. For example, constructing an artificial reef for a fish habitat with clean concrete rubble from razed buildings, sidewalks, and bridges does not require significant processing.

Recycle is the least preferred action under the pollution prevention hierarchy because it requires significant processing, which requires energy and produces pollution, although less than production from raw materials (e.g., collecting used paper to be recycled requires major processing and energy).

REFERENCES

U.S. Environmental Protection Agency (US EPA). 2016. Advancing Sustainable Materials Management: 2014 Fact Sheet. Available at https://www.epa.gov/sites/production/files/2016-11/documents/2014_smmfactsheet_508.pdf (verified 2 November 2017).

U.S. Environmental Protection Agency (US EPA). 2017. 2015 Biennial Report Summary Results for National. Available at https://rcrainfo.epa.gov/rcrainfoweb/action/modules/br/summary (verified 2 November 2017).

Environmental Modeling

INTRODUCTION

Modeling is a way to simulate or re-create reality. An **environment model** is a tool specifically designed to simulate or re-create the environment or, more specifically, an environmental system. It is often easier and less expensive to work with models than the actual system. However, models are valuable only if they are properly constructed and are fed good data; the popular saying "garbage in, garbage out" applies to modeling.

Models are generally of two types: static and dynamic. **Static models** are used to understand the behavior of a system at rest. Economists use static models extensively. **Dynamic models** allow us to examine a system over time and are used by environmental scientists to examine changes in an ecosystem.

This problem set provides you with the opportunity to explain the basic concepts of modeling and use a model to make determinations about an environmental system. It should help you to be able to describe several major challenges facing environmental regulators.

TASKS

Stock and flow modeling is the most basic form of dynamic environmental modeling. As shown in Figure P13.1, an example of a stock-and-flow model is a human population. You have births and immigrants flowing in (inflow), a population (stock), and deaths and emigrants flowing out (outflow).

Based on the information below, you will be modeling the level of a particular contaminant in a pond and answering a series of questions based on the use of the model.

Scenario:

- The Copper Brothers Manufacturing Company is located on the western shore of Valley Pond as shown in Figure P13.2.

- The pond's volume is 4×10^7 m^3 of water.

- The average flow-through rate is 8×10^4 m^3/day, that is, (1) the inflow from Little Valley Stream, (2) the water being discharged from the company into the pond, and (3) the feeder springs that collectively equal the volume of the outflow in Big Valley Stream (i.e., inflow – outflow) at 8×10^4 m^3/day.

- The company produces decorative copper art by chemically etching the copper with strong sulfuric acid.

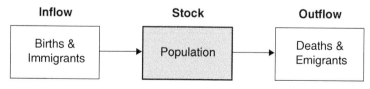

FIGURE P13.1 Simple stock-and-flow model

FIGURE P13.2 Valley pond study area

- The plant has a National Pollutant Discharge Elimination System (NPDES) permit issued by the state under the Clean Water Act. The NPDES permit allows the plant to discharge 0.16 metric tons of copper sulfate per day. The plant has an exemplary record of permit compliance. Currently, 25 people are employed at Copper Brothers. The company is the primary employment base for the town of Valley View.

- A family purchased a small camp on the eastern shore of Valley Pond. In the summer, on numerous occasions, they observed dead fish in their tiny cove near Big Valley Stream. They contacted the state Water Quality Division to file a formal complaint against Copper Brothers.

We need to calculate the steady-state level of copper sulfate in Valley Pond to determine if there is too much concentration in the pond. The steady-state level refers to how much copper sulfate is in the pond on a regular basis as a result of inflows, outflows, and copper sulfate discharges. Although the plant is in compliance with its permit, the level of copper sulfate may be too high for certain fish species because it could be accumulating in the pond even if there is a drop in discharges of copper sulfate into the pond from external sources.

The rate at which copper sulfate is added to the lake is known (0.16 metric tons per day). So, to calculate the steady-state stock of the pollutant, we need to know its residence time in the pond.

We will assume that the pollutant is uniformly mixed in the pond and is highly water soluble. (As with all models, certain assumptions must be made.) Thus, the residence time of the pollutant is equal to that of the pond water. We can calculate the residence time of the pond water as follows:

$$\textit{Residence time}: \mathrm{T}_w = \mathrm{M}_w / \mathrm{F}_w$$

T_w = residence time of water in the pond

M_w = stock of water: the pond volume

F_w = average, daily flow-through rate of the water

1. What is the residence time of the pond water? Write your equation with units, plug the numbers in, and show all work.

   The steady-state stock of copper sulfate can be calculated based on the following formula:

   $$\text{Steady stock: } S_{cs} = F_{cs} \times T_{cs}$$

   S_{cs} = steady-state stock of copper sulfate
   F_{cs} = daily discharge amount of copper sulfate
   T_{cs} = residence time of copper sulfate (see T_w and assume $T_w = T_{cs}$)

2. What is the steady-state stock (load) of copper sulfate in Valley Pond?

3. What is copper sulfate? And what are its likely effects on Valley Pond?

   The state's environmental standard for copper sulfate in aquatic systems is 1.98 parts per million. That is, 1.98 parts of copper sulfate per one million parts per water are allowed. We need to calculate the concentration of the copper sulfate in the water. This rather simple calculation requires us to divide the steady-state stock of copper sulfate by the total volume of Valley Pond (then multiply the number by 1×10^6):

   $$C_{cs} = S_{cs}/M_w$$

4. What is the concentration of copper sulfate in Valley Pond (expressed in ppm)?

5. Based on the concentration, as a state environmental scientist, what would you communicate to the Chief of the Water Quality Division about the effect of copper sulfate on water quality?

   The Copper Brothers Manufacturing Company asserts that if it is prevented from discharging copper sulfate into Valley Pond, it will have to close, which will put all 25 employees out of work. The company hires an engineering consultant to investigate manufacturing alternatives. The consultant concludes that Copper Brothers cannot reduce the quantity of copper sulfate it discharges without adversely affecting product quality. The company hires a consulting environmental engineer to present an ecological alternative, which the company proposes to the state. The company proposes to construct a pipeline that would connect Mountain Pond to Valley Pond, which, they argue, would dilute the copper sulfate below any adverse level. Essentially, the project would drain Mountain Pond and increase the volume of Valley Pond. Thus, as designed, the pipeline would increase the volume of Valley Pond by 1 million m³ (1×10^6 m³). However, the flow-through rates (F_w) would not change. Using this assumption, answer the following:

6. What is the revised residence time of copper sulfate?

7. What is the revised, predicted steady-state stock (load) of copper sulfate in Valley Pond?

8. What is the recalculated concentration (ppm) of copper sulfate? Will this proposal meet the state's water quality standard?

9. What are some likely environmental effects (i.e., unintended consequences) of increasing the water volume of Valley Pond by using Mountain Pond?

10. Is this proposal likely to appease the owners of the camp? Why or why not?

    You ask a colleague to review the consultant's report. Your colleague notes that there is a glaring error in the model's assumptions—the consultant did not take into account evaporation! Evaporation would have a significant impact on copper sulfate concentration because evaporating water contains no copper sulfate. Thus, the steady-state concentration of copper sulfate would be expected to be significantly higher because one of the possible exit pathways (evaporation) is no longer available. (That is, the flow-through rate and thus the residence time are wrong.)

Therefore, the residence time of copper sulfate is no longer equal to that of the water, but rather the residence time is associated only with the outflow of Big Valley Stream. Based on some rough calculations, the total rate at which water exits Valley Pond is 33% by evaporation and 66% through Big Valley Stream.

11. What is the revised residence time of the copper sulfate *without* considering the proposed pipeline? (Remember that the original flow-through rate is reduced by 33%.)

12. What is the revised, predicted steady-state stock (load) of copper sulfate in Valley Pond?

13. What is the new concentration (ppm) of copper sulfate? How does this amount relate to the state's water quality standard?

The state is concerned that this modeling approach may not be sufficiently accurate. The state regulators ask for general comments regarding the model:

14. What are some additional inflows and outflows that should be considered?

15. Suggest some additional relevant factors that were not incorporated into the model (e.g., review the assumptions). Think of factors that affect the pollutants and the living organisms.

16. What are some environmentally safer alternatives that Copper Brothers could employ?

17. As a state environmental regulator, what are some major challenges that you would face in determining levels of pollution in waterbodies using models?

Environmental Awareness and Ecological Identity

A society that knows its natural environment is better able to understand and solve environmental problems. Everyone has some level of environmental awareness even if they do not consciously think they do; everyone has an ecological identity of some sort or another. Your ecological identity arises from your awareness of the environment and is influenced by your life experiences, personality and values.

   We start with environmental awareness—the first stage is learning more about the environment in general and, in particular, environmental problems and their solutions. Awareness is simply perceiving or sensing the environment around you; knowing it is there. It begins with observing environmental phenomena. Observation is an important component of environmental science and is the first step in the scientific method.

ENVIRONMENTAL AWARENESS

1. Go to a designated spot outside of the classroom building or your home.
   A. Where are you—are you in a city, a college campus, a rural area, a forest—describe the setting? What are the general ambient (background) weather conditions?
   B. How windy is it? (Estimate the wind speed in miles per hour.) There are lots of ways to estimate wind speed. You could simply drop a piece of fuzz to see if it drops or drifts or you could look at trees, smoke, or a flag. Smoke rises straight up, and a flag droops straight down, if there is no wind.
         If the wind is between 1 and 4 mi per hour (mph), the flag only occasionally flips open and the outer end hangs lower; smoke drifts to the side; between 4 and 8 mph (a light breeze), the flag stirs more noticeably; and 8 to 13 mph is a gentle breeze that moves branches. See if you can determine a reasonable estimate by looking at clues. What method did you use to make your estimate?
   C. From which direction is the wind blowing? If you have a compass or a compass app on your phone, provide the direction in degrees.
   D. Are there clouds? If so, what formed them? What kind of clouds do you think they are (e.g., altocumulus, nimbostratus)? What methods could you use to estimate their height from the ground? How high would you guess them to be?

2. What is the temperature? Be sure to specify Celsius or Fahrenheit. Is it particularly warm or cold for this time of year and time of day? What is the humidity? Is it dry, moist, or muggy?

3. Look around your site. What signs of wildlife do you see?

4. Are there trees on your site? If so, what types? Is there other vegetation? If so, describe it.

5. Find a spot to sit in for 10 minutes and observe your surroundings, listening, looking, smelling, feeling, and sensing. This is to be a more focused observation than you did for item 1.B; the idea is to pay really close attention to what your senses are telling you about where you are.
   A. What do you hear?
   B. Glance around. What do you see?
   C. Form your two thumbs and index fingers into a circle (about 10 cm in diameter), and place it on the ground. Then count how many different species (plant and insect) you see in the circle. How many did you see? Were these more or less than you expected?

D. What do you smell?

E. What do you feel around you—are there hard surfaces, grass, or dirt?

F. What do your overall senses tell you about this place? What are your thoughts upon taking some time to observe and "feel" the place you are in?

ECOLOGICAL IDENTITY

Now take the time to think about your own sense of ecological identity. An *ecological identity* is the sum of the various ways that you see yourself in relationship to Earth—how it is reflected in your personality, actions, values, and behavior. It refers to your connection to Earth, your understanding of ecosystems, and your past experiences with nature, which collectively contribute to your sense of self (Thomashow, 1995). Ecological identity is the process of constructing meaning and an ecological worldview that can promote personal change. You have an ecological identity, but you just may not have thought about it before.

6. How have your awareness, knowledge, and sense of the environment contributed to your own ecological identity? In a short essay, describe your personal ecological identity and how it influences what environments have particular meaning to you, and how you live your life in terms of resource use and appreciation of nature.

7. Environmental awareness is one component of ecological identity. Imagine you are an environmental educator preparing to work with a group of middle school students in your community. Create a list of 10 questions that you think would promote environmental awareness about a place in that community that has meaning to you—it can be any setting; a park, street, corner lot, pond, wetland, a tree, tribal land. Select questions that are not just about facts, but rather use questions that seem important for awareness and that could contribute to ecological identity. For the questions that are not open ended (e.g., *What colors do you see?*), provide the answers and their sources.

REFERENCE

Thomashow, M. 1995. Ecological Identity: Becoming a Reflective Environmentalist. The MIT Press, Cambridge, MA.

Trophic Levels and a Tidal Marsh

Tidal marshes are highly productive ecosystems located at the interface between freshwater and marine environments. As an important habitat for a suite of organisms, salt marshes play an essential role in the life cycle of a large number of migrating fish and birds with regional economic and recreational significance.

Tidal marshes form in low-lying coastal areas that are protected from excessive winds, waves, and currents. Plant communities develop on the sediment base. Once plants are established, they trap additional sediments, and the increased deposition of sediments raises the marsh elevation, allowing expansion of the tidal marsh. Throughout this period, salt marshes have maintained themselves at the tidal elevations necessary for plant growth through the accretion of sediments filtered from tidal waters and the formation of peat from plant fragments and sediments. Two common Atlantic coast species are Smooth Cordgrass (*Spartina alterniflora*), which grows in areas flooded by daily tides, and Salt Meadow Grass (*Spartina patens*), which grows at a higher elevation in irregularly flooded areas.

Despite strong state and federal wetland laws that have virtually eliminated most filling and draining of tidal marshes, many factors continue to threaten the long-term ecological integrity of these ecosystems. Typically located in areas of concentrated coastal development, tidal marshes are threatened by fragmentation of the upland fringe, siltation and polluted runoff, rising sea levels, invasive plant species such as common reed (*Phragmites australis*), and legacies of past activities, including ditching, fragmentation by roads, and restriction of tidal flows. All of this and rising sea levels call increasing attention to salt marshes.

Picture a large, 3,000-acre salt marsh[1] as a critical habitat for a broad array of wildlife, particularly birds such as raptors, ducks, geese, egrets, herons, glossy ibises, sandpipers, willets, and grassland sparrows. Muskrat, mink, otter, and deer also frequent the marsh. The marsh has important ecological value in controlling stormwater, filtering pollution, and providing habitat. Most commercial shellfish and finfish depend on coastal wetlands for nursery and breeding habitat or on forage fish that breed in coastal wetlands.

On any given day, sinuous-necked Great Blue Herons wade in pools alongside Snowy Egrets in search of Mummichugs and Sticklebacks. Glossy Ibises probe the mud for worms with their long curved bills, Willets and Yellowlegs skitter the edges of panes (shallow ponds separate from the tidal stream). Cormorants perch on driftwood and spread their wings, like swimmers hanging beach towels to dry. Ribbed mussels, amphipods, snails, soft-shelled crabs, American eel, and Killifish swim, crawl, or burrow in the marsh's waters and mud.

Common name	Scientific name	Habitat
Smooth cordgrass, tall form	*Spartina alterniflora*	Low marsh
Smooth cordgrass, short form	*Spartina alterniflora*	Poorly drained high marsh
Salt meadow grass	*Spartina patens*	High marsh
Black grass	*Juncus gerardii*	High marsh
Common reed	*Phragmites australis*	Brackish marsh

---

[1] We will use the northeastern United States for the purposes of selecting representative species in the marsh.

Common name	Scientific name	Habitat
Narrow-leaf cattail	*Typha angustifolia*	Brackish marsh
Salt marsh bulrush	*Scirpus spp.*	Brackish marsh
Spike grass	*Distichlis spicata*	High marsh
Seaside milkwort	*Glaux maritima*	High marsh
Broad-leaf cattail	*Typha latifolia*	Fresh/brackish marsh

ADDRESS THE FOLLOWING ITEMS USING THE INFORMATION ABOVE, INTERNET OR LIBRARY RESEARCH, AND YOUR TEXTBOOK:

1. Which of the following species are in the phylum Chordata?

     Smooth cordgrass, salt meadow grass, common reed, salt marsh bulrush, broad-leaf cattail, spike grass, seaside milkwort, black grass, plankton, clam worms, ribbed mussels, amphipods, salt marsh snail, fiddler crab, blue crab, killifish, mummichugs, stickleback, willets, yellowlegs, cormorants, clapper rail, snowy egret, glossy ibis, grassland sparrow, laughing gull, herring gull, ducks, geese, cattle egret, great blue heron, sandpiper, mice, shrew, muskrat, raccoon, mink, otter, deer, and fox.

2. Which of the following species share a genus?

     Smooth cordgrass, salt meadow grass, common reed, salt marsh bulrush, broad-leaf cattail, spike grass, seaside milkwort, black grass, plankton, clam worms, ribbed mussels, amphipods, salt marsh snail, fiddler crab, blue crab, killifish, mummichugs, stickleback, willets, yellowlegs, cormorants, clapper rail, snowy egret, glossy ibis, grassland sparrow, laughing gull, herring gull, ducks, geese, cattle egret, great blue heron, sandpiper, mice, shrew, muskrat, raccoon, mink, otter, deer, and fox.

3. Trophic levels are a way of portraying relationships in a food web (Figure P15.1). The lower the level the more biomass and the greater the number of individuals; thus the levels and their occupants may be represented on a triangle (Figure P15.2). Place 25 of the previous species on the trophic triangle. Put them at the level you think they belong, making sure

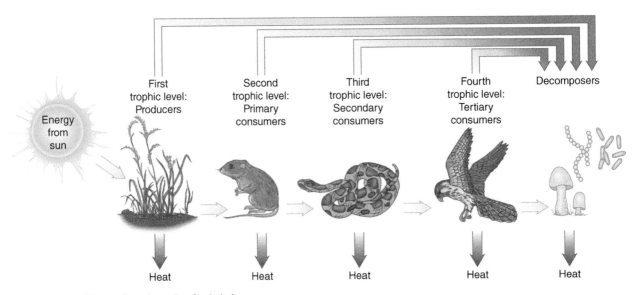

FIGURE P15.1   Energy flow through a food chain

Source: Raven, P.H., L.R. Berg, and D.M. Hassenzahl. 2008. Environment. 6th ed. Wiley & Sons, New York.

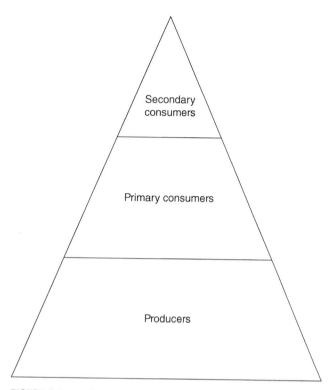

FIGURE P15.2   Trophic triangle with the first three trophic levels

that there are some in each of the three trophic levels. Use arrows to show energy flow links among the species you selected.

4. Estimate the biomass at each trophic level for 1 acre. Write your assumptions and how you arrived at them. For example, you can assume one fox, one owl, and one heron per acre. Make other assumptions about weights and the number of individuals per acre for these top trophic species. Assume that each trophic level requires 100 times its biomass for the trophic level below it. Start at the top and work down.

Food Efficiency: The Breakfast Assessment

What we eat has a direct impact on the environment. As illustrated in Figure P16.1, the production, processing, storage, and distribution of food requires energy, pesticides, fertilizers, and a variety of other inputs. Moreover, there are significant inputs and outputs in packaging, selling, and advertising food products. The energy needed to provide food can be estimated as a function of growing food, processing it, and transporting it to the market. If the market reflects the true cost of production, then the market cost (i.e., retail price) is a good representation of energy, and even if it is not, it still represents the energy that it took for you to earn dollars. Food returns energy to us in the form of calories (a food **calorie** as used in product labeling and nutrition is actually the same as a **kilocalorie** as used in science, which is the amount of energy required to raise the temperature of 1 L of water 1°C at sea level). Nutritional value represents efficient calories that provide what your body needs to stay active and healthy.

Land that produces plant products yields more kilocalories per square meter per year than land used to produce animal products. Animals are consumers, and as energy flows up from lower trophic level producers (plants), there is a loss of energy between the trophic levels through respiration, heat, and animal waste.

Plants can be divided into two groups based on their photosynthetic and respiratory pathways. C-4 plants are fast, efficient photosynthesizers compared to C-3 plants.[1] Cane sugar is a C-4 plant with a yield of 3,500 kcal/m²/year; corn cereal—another C-4[2]—yields 1,600 kcal/m²/year. Tables P16.1 and P16.2 show the yields of certain plant and animal food sources (Brewer and McCann, 1982). Suppose that a slice of wholegrain bread is estimated at 65 cal. Table P16.1

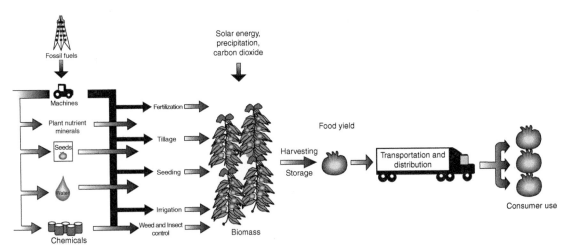

FIGURE P16.1 Inputs/outputs of food production, processing, transportation, and distribution

---

[1] Most plant species are C-3 (or C3), they are called so because they make a three-carbon compound as a stable product of carbon fixation. Plants that make a four-carbon product with PEP carboxylase are called C-4 (or C4) and lose far less carbon through photosynthesis than do C-3 plants. Less than 1% of plant species are C-4.

[2] 10,000 square meters equal 1 hectare or 2.47 acres. Caloric yields are just estimates for comparative purposes since many factors affect volumetric production as well as food quality. Data for Tables P16.1 and P16.2 are based on Akoroda (1998); Brewer and McCann (1982); Smil (2001); U.S. Department of Agriculture, National Agricultural Statistics Service (2004); and Waggoner (1994).

Table P16.1   C-3 Plant Categories and Their Yield in kcal/m²/year

Bread	650	Apples	1,500
Wheat (and cereal)	810	Pears, peaches	900
Oranges, grapefruit	1,000	Vegetable oil	300
Frozen orange juice	410	Margarine	300
Peanut butter	920	Beet sugar	1,990
Rice (and& cereal)	1,250	Coffee	12
Potatoes	1,600	Tea	40
Other vegetables	200		

Table P16.2   Animal Products and Their Yield in kcal/m²/year

Milk	420	Beef	130
Eggs	200	Cheese	40
Chicken	190	Fish	2
Pork	190		

indicates that 1 m² is needed to grow sufficient grain for 650 cal (sufficient for 10 slices of that bread); therefore, the amount of land needed for the slice of bread is a tenth of a square meter. Two slices would be 0.2 m².

1. For yourself or another person, assess a recent breakfast meal. Write the food items in the left column of Table P16.3.
   - The USDA has a food tracker to assess calories, sugar, saturated fat, and sodium content for many foods. (Remember: 1 cal is same as 1 kcal.) Visit the food tracker at https://www.supertracker.usda.gov. Use it to help you fill out the breakfast assessment table below.
   - What portion of the average daily amount of nutrition would you estimate is provided by the breakfast?

Table P16.3   The Breakfast Assessment

Your name: _____ Date: _____ Person assessed: __self, __other: _____							
Food item and estimated calories	Calorie	Low, moderate, or highly processed	Miles driven to purchase this food	Distance food traveled from origin, 0–200, 200–1,000, or >1,000 km	Market cost in $	Square meters to grow	Sugar
Total			Avg =				

Note: Make a key or guide that provides your interpretation of the column headings and the assumptions that you made in completing the chart.

2. In completing the table, you will need to determine if the food represents a high, moderate, or low degree of processing (processing refers to mechanical or chemical operations on food necessary to change or preserve it sufficiently for the consumption). This should be based on the transformation the food source undergoes in manufacturing and on the amount and type of packaging. Highly processed food such as a candy bar, despite its high calories, tends to have low nutritional value. A bowl of cereal might have the same calories, but generally is better for you in terms of nutrition. If a meal meets one third of a person's daily need for nutrients, consider it high in nutritional value.

3. Assume that the efficiency of the breakfast is a function of the land area to grow the food, degree of processing, energy expended in driving to get the food, distance the food traveled, market cost, packaging, and nutritional value. How efficient was the person's breakfast? Explain your reasoning for your answer.

4. How does it compare with what a poor person in a developing nation would typically eat? Cite any references you use in support of your answer.

5. How does the breakfast compare with what a middle-class person in northern Europe would typically eat? (Use your textbook or the Internet to help you.)

6. Carbon footprint calculators can be used to determine CO_2 emissions for food. For example, a typical serving of beef yields 6.1 lb of CO_2, an 8 oz glass of milk produces 0.72 lb, and a glass of fruit juice produces 0.54 lb (Heller and Keoleian, 2014). Find a carbon footprint calculator and use it to help you estimate the CO_2 emitted by the meal you assessed. Be sure to cite your references.

REFERENCES

Akoroda, M.O. 1998. Comparative Output of Calories from Starchy Food Crops in Sub-Saharan Africa. Tropical Agriculture. 75:257–262.

Brewer, R., and M.T. McCann. 1982. Laboratory and Field Manual of Ecology. Saunders College Publishing, Philadelphia, PA.

Heller, M., and G. Keoleian. 2014. Greenhouse Gas Emissions Estimates of U.S. dietary Choices and Food Loss. Journal of Industrial Ecology. 19 (3): 391–401.

Smil, V. 2001. Enriching the Earth: Fritz Haber, Carl Bosch, and the Transformation of World Food. MIT Press, Cambridge, MA.

U.S. Department of Agriculture. 2017. SuperTracker. Available at https://www.supertracker.usda.gov/ (verified 8 November 2017).

U.S. Department of Agriculture, National Agricultural Statistics Service. 2004. Agricultural Statistics: An Annual Report. Available at https://www.nass.usda.gov/Publications/Ag_Statistics/ (verified 8 November 2017).

Waggoner, P.E. 1994. How Much Land Can Ten Billion People Spare for Nature? Council for Agricultural Science and Technology, Ames, IA.

Life-Cycle Assessment

The manufacture, use, and/or disposal of all products have some impact on the environment. However, some products require more nonrenewable resources, use more toxic materials, have to travel further, require more energy for proper storage, emit more pollution, or cause greater impacts when used and then ultimately disposed. To promote sustainability, we need to design, produce, and use products that require less resources, energy, transportation, and toxic materials. **Design for the Environment** (DfE) is the design of products and systems that incorporates pollution prevention and resources and energy conservation by systematically considering the health and environmental impacts throughout a product's life cycle. We should produce products that have a minimal impact when reclaimed and that can be more easily reused or recycled after their useful life through a process referred to as design for recycling or demanufacturing, which focuses on a product's end of life.

One approach to improving the sustainability of products is to assess its life cycle. **Life-cycle assessment** (LCA) is a process for identifying and evaluating the potential environmental effects of a product over its lifetime (US EPA, 1993). That is, a "cradle-to-grave" analysis. Because you generally focus on a product, which is the end result of a manufacturing/processing effort, you need to work backward—**upstream**—to examine the various environmental impacts resulting from every activity required to produce, distribute, use, and eliminate the product. LCA is a powerful tool that can be used to assist public agencies in formulating environmental policies, help manufacturers reduce energy consumption and the environmental impacts of their products, and help consumers make more informed choices.

To apply LCA, you must sketch the product's life cycle (from cradle to grave), identify resource inputs, identify outputs, assess the associated potential environmental impacts, and identify *pollution prevention* measures. (See side bar).

Conducting a complete and thorough LCA is complex, time-consuming, and expensive, as it requires detailed data and knowledge. However, a simplified, conceptual LCA can be produced relatively easily and provides important, basic information.

There are five major steps in applying LCA.

STEP 1: Create a flow diagram depicting the various steps for a product. (See Figure P17.1 for an example of a conceptual LCA depicting the life cycle, inputs, and outputs.)

– Raw material acquisition

– Manufacture

– Processing

– Use

– End-of-life management

STEP 2: Identify the major inputs: renewable and nonrenewable raw materials and types of raw energy (e.g., fuels).

POLLUTION PREVENTION

Pollution prevention is based on *source reduction*, which means any practice that (a) reduces the amount of a pollutant entering any waste stream or otherwise released into the environment prior to recycling, treatment, or disposal; and (b) reduces the impact on public health and the environment from the release of such pollutants. Source reduction practices include equipment or technology modifications, process or procedure modifications, reformulation or redesign or products, substitution of raw materials, and improvements in housekeeping, maintenance, training, or inventory control.

Pollution Prevention Act of 1990, United States Code Title 42, Chapter 133, Sec. 13102.5(A).

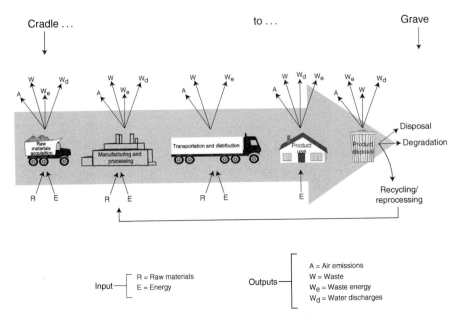

FIGURE P17.1 Conceptual life-cycle assessment with inputs and outputs

Assessing the life cycle, from cradle to grave, of a product is becoming more common in manufacturing and other corporate and government processes.

STEP 3: Identify the major outputs (e.g., the useable product, air emissions, water discharges, waste energy, and solid waste) including the specific type of pollutant and waste.

STEP 4: Evaluate the potential environmental impacts at each stage. Based on the inputs and outputs, the potential impacts on human health and the environment can be determined.

For example, if the primary fuel source is coal, the environmental effects may include erosion/sedimentation and acid mine drainage from mining; production of acid deposition, mercury emissions, and greenhouse gases through combustion; and landfilling of resultant coal-combustion ash. The manufacture and processing of the product may release volatile organic compounds, which are a principal component of ground-level ozone. The use of a product (e.g., automobile) may generate air emissions and result in spills to water and soil. The disposal of the product, when it is no longer wanted, may result in air emissions if it is incinerated, or releases to the soil and groundwater if it is landfilled (e.g., mercury battery).

STEP 5: Identify pollution prevention options for each energy source and environmental impact. For example, is there a cleaner energy alternative that can be used for the product? Can the product be manufactured with safer and less polluting substances? Can the produce use recycled materials in place of virgin raw materials? Can the product be designed with less packaging? Can the product be designed so that its emissions during use are reduced? Can the product be designed to increase its reuse or recycling potential so that landfilling or incineration can be avoided and the waste becomes a resource?

TASKS

- Select a consumer product with the help of your lab instructor.

- Create a poster that depicts and documents the principles of LCA for your product. Your poster should be neat and professional in appearance and reflect effective communication principles.

  1. Provide a detailed description of the product (i.e., dimensions, contents, use, and so forth). Include a photograph, drawing, or actual product. What is the mean volume or weight of the

product? How many/much are produced annually? What are the geographic sources of the major raw materials? Where is the product manufactured? Where are the primary markets? What are the main alternatives (e.g., paper bag vs. cloth or plastic, bamboo chopsticks vs. plastic ones)?

2. Using the Internet and library for research data, draw the product's conceptual life cycle on the poster as depicted in Figure P17.1 using a block flow diagram.[1] The life cycle should include the raw materials acquired, manufacturing/processing, transportation/distribution, disposal, and the environmental fate of the product and byproducts so you've captured the "cradle-to-grave" aspects.

   A. For *each* product stage, identify and list the input and outputs: For these steps, you will need to conduct some basic library or Internet research.
      I. List the likely energy source and raw materials used to produce the product (including its packaging) and the origin of the energy sources (e.g., if coal, where was the coal likely mined?).
      II. Where feasible, provide numerical data (e.g., x% aluminum, x% PVC plastic)
      III. List the likely emissions, types, and amounts of pollutants and wastes.

   B. For *each* product stage, identify and list the pollutants and environmental impacts (e.g., decreased water quality, increased ground-level ozone, soil degradation, CO_2 emissions, loss of habitat).

   C. For *each* product stage, identify possible pollution prevention measures that could reduce the overall environmental impact of the product, including production, transportation, use, and disposal.

   D. Be sure to cite references properly and clearly on the poster.

U.S. Environmental Protection Agency (US EPA). 1993. Life Cycle Assessment: Inventory Guidelines and Principles (No. EPA/600/R-92/245). Government Printing Office, Washington, DC.

REFERENCES

---

[1] A **block flow diagram** is a schematic illustration of a major process. The blocks (or rectangles) represent a unit operation or step. These blocks are connected by lines or arrows that represent the process flow between the unit operations/steps. Many figures in this manual are block flow diagrams.

Understanding Pesticide Labels

This problem set provides you practice in understanding pesticide labels and thinking about the need for accountability in product labeling for good consumer decisions regarding the environment.

Pesticides are not just for commercial agriculture. They are part of many things, including home maintenance, disinfection, lawn care, gardening, and insect repellants. Pesticide products exist all around us, even when we cannot see them. For this problem set, select a pesticide product from your home or go to a local store and find a pesticide product. Use the product's label[1] and Internet research to answer the following:

1. In what type of store did you find it or did you find it at home? If you found it in your home, where specifically was it stored? How accessible was the product to children?

2. What is the trade name of the product?

3. Where and how is this product supposed to be used?

4. What is the company that manufactured this product?

5. Discuss the label. Is the label clear and easy to understand? Is the print tiny?

6. What is the product's U.S. EPA registration number? What is the purpose of the registration number?

7. What is the name and amount of active ingredients? Research the active ingredients and describe any toxic effects that they might have on the environment, animals, and humans. Try to find objective references. Cite your sources.

8. What is the name and amount of inactive ingredients? Research these and describe any toxic effects that they might have on the environment, animals, and humans. Try to find objective references. Cite your sources.

9. What health and environmental warnings are listed on the container?

10. Would you use this product? If so, under what conditions would you use it or have you used it?

11. How safe do you think it is compared to similar products on the market?

12. Now that you have researched the active and inactive ingredients, evaluate the label's effectiveness in including information that you consider important.

13. Think about the potential relationship between your product and our food supply and discuss how the pesticide may be a contaminant.

14. What are your thoughts about the product after doing this problem set?

---

[1] You can also search the U.S. Environmental Protection Agency's database of registered pesticide labels, Pesticide Product and Label System, which can be searched by product name: https://iaspub.epa.gov/apex/pesticides/f?p=PPLS:1

Review and Reflection

This problem set reviews the main points of the previous problem sets; examines how you, as an individual, matter; and prompts you to look at the big picture.

Throughout this course, your environmental awareness was cultivated as you explored where and how you fit into the environment around you. The first two problem sets focused on the scientific method and its application in environmental science. From there, problem sets focused on how to quantify environmental problems and the impact of humans on environmental quality, in particular, population, resource use, pollution, and other activities that can negatively impact the environment. Throughout the course, the problem sets have encouraged you to examine your role in the environment. What are the consequences of the choices you make? Or the way you choose to live your life? Look back over your problem sets and think about how they all fit together.

WHAT DOES IT ALL MEAN?

Summarize each of the problem sets you completed and explain how they fit into the basic model of environmental degradation (ED): $ED = P \times A \times T$, where ED is environmental degradation, P is population, A (affluence) is per capita resource use, and T (technology) is environmental degradation per unit resource use.[1] Then describe how the problem sets and laboratory exercises that you completed have affected your understanding of environmental science from personal and academic perspectives.

SHORT ESSAY

---

[1] This model is a slight variation of the environmental impact model developed by Ehrlich, P.R., and J.P. Holdren. 1971. Impact of Population Growth. Science. 171:1212–1217.

Appendices

PART 4

Glossary

Abiotic factors Nonliving or physical factors (temperature, light) in the environment.

Adjacent Bordering or near. May be on a separate land parcel or lot. The determination of adjacency is often based on the potential for environmental impacts.

Air exchange rate Rate at which outdoor air replaces indoor air in a building or room.

Air pollution Contamination of the air through the release of harmful substances.

Alternative energy Energy that is not generated from fossil fuels.

Ambient air quality Overall condition of the outside air (compared to indoor air).

Anaerobic digestion The use of microorganisms to break down biodegradable material in the absence of oxygen.

Anthropocentric A human-centered world view.

Anthropogenic Human-made pollutants.

Aquifer An underground layer of permeable rock, sediment (usually sand or gravel), or soil that yields water.

Ash fill Landfills devoted to disposal of municipal solid waste incinerator ash.

Best management practices Practices, planning, policies, and engineered devices and systems to prevent, reduce, control, or treat contaminated stormwater.

Biodiversity Different species and life-sustaining processes that can best survive the variety of conditions found on the earth.

Biotic factors Living factors in the environment.

Block-face One side of a street between two intersecting streets or other geographic boundaries.

British Thermal Unit (BTU) Amount of energy required to raise the temperature of 1 lb of water 1°F when the water is near 39.2°F.

Brownfield Abandoned, idled, or underused industrial or commercial facility or site where expansion or redevelopment is contemplated in the presence of actual or potential contamination.

Caloric content Amount of food calories in a given food item.

Carbon dioxide measurements Term used for the measurement of the quantity of carbon dioxide in the air. Levels above 1,000 ppm can indicate inadequate ventilation indoors.

Carbon dioxide production Amount of carbon dioxide produced; the average gallon of gas produces 9 kg of CO_2.

Carbon footprint The total amount of greenhouse gases produced to directly and indirectly support human activities.

Carrying capacity Maximum abundance of a population that can be maintained by a habitat or ecosystem without degrading the habitat or the ecosystem.

Carson, Rachel Author of *Silent Spring* and an environmentalist who opposed the unrestricted use of pesticides, particularly DDT.

Clinometer Instrument for measuring angles of inclination. It can be used to determine slopes as well as the height of trees, buildings, clouds, and other features.

Combined sewer system (CSO) A sewage collection system that is designed to collect sewage, stormwater, and industrial wastewater in the same pipe.

Concentration–response curve Method used to document the effects of various concentrations of a chemical substance on a group of organisms.

Confounding variable An extraneous (uncontrolled) variable that could produce an alternative explanation for the results of an experiment.

Contamination The tainting of an item (i.e., soil) through human activities, including the intentional and unintentional discharge of hazardous materials and waste.

DDT (dichlorodiphenyltrichloroethane) An insecticide that is toxic to animals and humans, which was used mostly during the 1950s and 1960s, but is now banned due to its detrimental effects.

Degree day (DD) Degree days are used as an index originally used to calculate the energy to heat or cool a home, based on normal inside temperatures and averaged outside temperatures, but has come to be a standard reference in weather reports. Each degree of temperature below of 65°F is counted as one heating degree day (HDD) and each degree above 65°F is counted as one cooling degree day (CDD).

Design for the environment (DfE) The design of products and systems that incorporates pollution prevention and the conservation of resources and energy by systematically considering the health and environmental impacts throughout a product's life cycle.

Dissolved oxygen (DO) measurement Term used for the measurement of the amount of oxygen dissolved in a unit volume of water.

Drainage swale Depression or shallow channel between slopes that provides for the flow of surface or shallow groundwater away from an area. A swale may also be a low area of moist land.

Earth disturbance Land alteration activity or effect, whether intentional, accidental, or natural. Examples include erosion, excavation, creation of berms, construction-related impacts, and the effects of flooding.

Ecological diversity Variety of habitats (forests, grasslands, streams).

Ecological footprint The area of productive land required to provide resources and assimilate waste products to meet our consumption needs.

Ecological identity The sum of the various ways that you see yourself in relationship to earth—how it is reflected in your personality, actions, values, and behavior.

Effluent Liquid discharged into receiving water.

Energy conservation Any behavior or technology that results in consuming less energy (e.g., turning lights off when leaving the room, installing motion-detector-based lighting, or using a sleep function on electronics).

Energy efficiency Using technology that requires less energy to perform the same function (e.g., using LED lights that require less energy instead of fluorescent or incandescent lights, driving hybrid cars, using a clothes line to dry clothes).

Energy investment in food production Amount of energy that goes into producing, processing, packaging, and shipping food.

Environmental audit Inspection of a designated area to determine its overall environmental health.

Environmental equity The fair distribution of land and resources between humans and other organisms.

Environmental impact An adverse effect on the environment. Alteration of landscape or environmental setting, usually to its detriment. Impacts may occur in a variety of categories, including water quality, soils, air quality, biodiversity, and aesthetics.

Environmental model A tool specifically designed to simulate or re-create the environment, or more specifically, an environmental system.

Environmental risk The ability of an agent (chemical, physical, or radioactive) to cause harm to human health and/or the environment.

Environmental toxicology The study and detection of environmental poisons and their effects on humans and the environment.

Exhaust emissions Fumes released from vehicles, including sulfur oxides, nitrogen oxides, volatile organic compounds, particulate matter, and carbon monoxide.

Experimental design The creation of an experiment through which a hypothesis can be tested. The experiment must adhere to the scientific method.

Exposure The potential for a person to come into contact with a contaminant.

Field screening techniques Techniques used in determining the source of an environmental problem, including specific conductance, dissolved oxygen, turbidity, ammonia, and pH.

Flood plains (floodplains) Flooding is a natural and recurring event for a river or stream. On average, streams will equal or exceed the mean annual flood level about once every 2 or 3 years. Floodplains are, in general, those lands most subject to recurring floods, usually situated adjacent to rivers and streams. Floodplains are therefore "flood-prone" and are hazardous to development activities if the vulnerability of those activities exceeds an acceptable level. Floods are usually described in terms of their statistical frequency. A "100-year flood" or "100-year floodplain" describes an event or an area subject to a 1% probability of a certain size flood occurring in any given year.

GIS (geographical information system) A combination of hardware, software, and procedures used to manage spatial (georeferenced) data.

Global environment The environment of the entire world.

Global sustainability Efficient use of resources to sustain the current population and allow for the use of resources for future generations.

GPS (global positioning system) A radio navigation system that allows land, sea, and airborne users to determine their exact location, velocity, and time 24 hours a day, in all weather conditions, anywhere in the world. GPS is used to support a broad range of military, commercial, and consumer applications.

Greenwashing When a corporation or organization deceptively uses the environment, including concepts, symbols, and slogans, to sell products or to sell a corporate or organizational persona when the product, corporation, or organization would not normally be viewed as "green" or sustainable.

Habitat Local environment in which an organism, population, or species lives.

Heat loss Loss of heat due to inefficient insulation.

Heliodon (HEE-leo-don) A device for adjusting the angle between a flat surface and a beam of light to match the angle between a horizontal plane at a specific latitude and the sun or other light source. Heliodon can be used to determine the amount of light that a feature (e.g., a skyscraper, house, or yard) could receive from the sun during a given day or period.

Hydraulic fracturing (fracking) Process by which a high-pressure fluid containing water, sand, and chemicals is injected into a well designed to create cracks in deep-rock formations as a means to extract natural gas and petroleum.

Hypothesis A tentative statement that proposes a possible explanation of some phenomenon or event.

Hypothesis testing Testing of a hypothesis through a designed experiment that follows the scientific method.

Indoor air quality The nature of air that affects the health and well-being of occupants.

Inductive reasoning Going from the specific to the general (opposite of deductive reasoning). Much of science involves observing specific phenomena and drawing general conclusions about them. Example: *"The sample of 20 flies of the same species is hermaphroditic; therefore, all other members of this species is hermaphroditic"* is an inductive statement that might not necessarily be true (hence care is taken in using the word "proof" in science).

Infiltration The process by which outdoor air flows into a building or room through openings, joints, and cracks in walls, floors, and ceilings, and around windows and doors.

Institutional Review Board (IRB) A generic term for a committee designated to approve, monitor, and review any research activity that involves living things, especially human subjects. The term IRB is used by the U.S. Food and Drug Administration (FDA), the Department of Health and Human Services (HHS), and other entities; however, equivalent reviewing bodies might also be known as an **independent ethics committee** (IEC) or an **ethical review board** (ERB).

Kilowatt-hours (kWh) Unit of energy equivalent to 1 kW of power expended in 1 hour; not a standard unit, but is commonly used in electrical applications.

Laboratory analysis The analysis of data through observation and experimentation in a laboratory.

Landfill A plot of land used for the long-term storage of solid waste.

Leachate Liquid that seeps through or from a landfill or other waste storage area.

Lethal concentration Concentration of a chemical that causes death to organisms; most commonly measured at LC_{50}—the concentration that kills half of test organisms. The lower the LC_{50} value, the greater the toxicity.

Lethal dose As opposed to concentration, it is the dose given directly to an organism, typically by injection or oral administration.

Liability A legal responsibility, duty, or obligation. The state of one who is bound in law and justice to do something that may be enforced by action. Similarly, concerns about liability mean that environmental investigators only enter land on which they have permission or authority.

Life-cycle assessment A process used to identify and evaluate the potential environmental effects of a product over its lifetime.

LOAEL (Lowest Observed Adverse Effect Level) The smallest tested dose of a substance that causes harmful effects on the tested subject (plant or animal).

Lumen A measure of light energy from a light source. A lumen is the light available in a square foot area located 12 in away from a single candle. A 100-W incandescent light bulb will put out approximately 1,500 to 1,700 lm.

Lux (lx) A unit of luminance (in SI notation). It is defined as lumens per square meter (lm/m^2).

Material Safety Data Sheet (MSDS) A government-required informational document prepared by the manufacturer that describes the physical, chemical, and hazardous properties of a product.

Mechanical ventilation The use of devices, such as outdoor-vented fans, to intermittently remove air from a room, such as bathroom, to air handling systems that use fans and duct work to continuously remove indoor air and conversely distribute filtered and conditioned outdoor air to strategic points throughout a building or house.

Metabolites Byproducts produced from metabolism—the organic processes necessary to sustain life.

Microclimate The climate of a small, specific place within an area as contrasted with the climate of the entire area.

Modeling A tool to simulate or re-create reality.

Municipal solid waste Solid waste that is nonhazardous and nonindustrial, which includes trash, garbage, rubbish, and refuse from residential, commercial, and institutional sources.

Natural ventilation Air that moves through opened windows, doors, and passive vents.

Nonrenewable energy Energy sources that cannot be replaced once they have been used.

Null hypothesis Hypothesis that states that the variables do not have an effect on the outcome of the experiment.

Nutritional value Amount of nutrients (vitamins, minerals) in a given food item.

Parent material Material from which a soil is formed.

Passive solar energy The capture of sunlight directly and its conversion to low-temperature heat used domestically for heating air and water; it does not use any mechanical devices.

Pedosphere The thin layer of soil on the earth.

Peer review A process used for checking and verifying the work performed by one's equals—peers—to ensure that it meets specific academic criteria.

Per capita energy consumption Total amount of energy consumed per person.

Personal accountability Recognizing and taking personal responsibility for the resources that the individual consumes.

pH A measure of the acidity or alkalinity of a solution, numerically equal to 7 for neutral solutions, which increases with increasing alkalinity and decreases with increasing acidity. The pH scale commonly in use ranges from 0 to 14.

Phase I Site Assessment The first phase of an environmental site inspection, including the basic observation of the environmental health of a designated area.

Phytotoxicity Poisonous to plants.

Plume Volume of contaminated groundwater that extends downward and outward from a specific source.

Point source A discharge into water from a discrete, fixed discharge point such as a pipe, culvert, ditch, or tunnel.

Pollution prevention The prevention of pollution through the reduction of waste generated, the reuse of otherwise disposable products, and the recycling of materials.

Pollution Prevention Act of 1990 A Congressional act that established a national policy that includes the prevention or reduction of pollution at the source wherever possible, the reuse and recycling of pollution in an environmentally safe manner whenever feasible, the treatment of pollution that cannot be prevented or recycled, and the disposal of pollution into the environment only as a last resort and in an environmentally safe manner.

Population A group of individuals of the same species living in the same area.

Population growth The growth of a population based on the number of births, deaths, immigration, and emigration.

Potable Water suitable for drinking.

Precautionary principle The guiding principle that when an activity has the potential to cause significant threats of harm to the environment and/or human health, we should act even if we have not yet fully identified or understood the cause-and-effect relationships.

Primary treatment The initial treatment applied, particularly regarding sewage treatment, which utilizes bacteria to degrade waste.

Range of tolerance The varying range of environmental conditions that a species can tolerate; individuals within a species may also have slightly different ranges of tolerance.

Raw data Data that has not been processed (i.e., extracted, organized, formatted, summarized, or analyzed) for presentation.

Recycle Reducing pollution by reprocessing materials into a new usable form.

Reduce The reduction of pollution by using fewer materials.

Regulated activity Construction or other land-uses, including discharges, storage, and movement of materials subject to federal, state, or local jurisdiction under an applicable law. Regulated activities are either allowable by general rule or by specific permit.

Remediate To improve the quality of something by repairing a deficiency. In environmental assessment, it is the treatment of a contaminated site by removing contaminated solids or liquids, or by treating them on-site.

Research techniques Procedures used in gathering information.

Reuse Reducing pollution by reusing materials such as containers or water in the cooling systems of facilities.

Risk perception An individual's assessment, based on feeling or judgment, of the potential for harm.

RMS voltage Voltages such as the common U.S. household standard of 120 V are actually listed as RMS or root mean square. RMS can be used for either voltage or current. It is achieved by taking the potential difference at each instant or sample point, squaring it, adding up these figures, and dividing by the number of samples to find the average square or mean square. The square root of this figure yields the RMS average value.

R-value Measurement of the ability of a material to resist heat transfer. An R-value of 1 is equal to the number of BTUs that would pass through a 1-ft^2 surface in 1 hour if the difference in temperature on opposite sides of the surface is 1°F.

Scientific method The process of experimentation, which includes making an assumption about an observation or a phenomenon, creating a hypothesis, testing the hypothesis, and accepting or rejecting the hypothesis based on the results of the experiment, and, finally, revising the hypothesis.

Secondary treatment The second treatment applied, particularly regarding sewage treatment, to ensure the degradation of waste before it is discharged.

Sedimentation basin A depression in the land (basin or pond) constructed to handle excess runoff and sediment from developed land parcels.

Sequential comparison index (SCI) The result of a biodiversity estimation method that conducts a rough picture of species diversity without having to identify the specific species.

SI *Le Système International d'Unités* (International System of Units) is the world's most widely used system of units. It was developed from the metric system in 1960 and is based on the powers of 10.

Sick building syndrome A set of symptoms that affect some occupants during time spent in a building and diminish or go away entirely during periods spent away from the building, but cannot be traced to specific pollutants or sources within the building.

Soil disturbance A disturbance in the soil such as burrowing or filling.

Soil horizon The layers that make up a soil profile. Each horizon can be identified by changes in color or texture.

Soil profile A cross-sectional view of soil.

Soil texture triangle The three variables (sand, silt, and clay) used in determining soil type based on texture.

Solar gain An increase in temperature due to the sun shining on a landscape, a room, or an object.

Species evenness Relative proportion of each species.

Species richness Number of different species.

Spirometry The measurement of breathing capacity, rate, and volume. Important in the maintenance of human health, especially regarding asthma and other respiratory illnesses.

Steady-state model Type of model in which the stock (such as fish in a pond) does not change or is balanced by equal inflow and outflow.

Stormwater runoff Runoff generated from precipitation and/or snowmelt events that flow over impervious surfaces, such as paved streets, sidewalks, parking lots, and roofs, or land.

Superfund Officially known as the Comprehensive Environmental Response, Compensation, and Liability Act, this program establishes strict liability for property owners where hazardous substances present a threat to human health or the environment, regardless of when the hazardous substances were placed there.

Survivorship curve A graphical representation of the likelihood that an individual will survive from birth to a particular age.

Testable Relating to hypotheses, there must be an ability to determine whether the results are logically possible (i.e., true or false).

Topographic map (topographical map) A map depicting terrain relief showing ground elevation, usually through either contour lines or spot elevations. The map depicts the horizontal and vertical positions of natural and human-made features. U.S. Geological Survey maps of 7.5 minutes and 15 minutes are commonly used for field navigation and may be provided as readouts from digital equipment (**GPS** units). More detailed site maps will be on a much larger scale, showing meter or foot increments in elevation.

Toxic release Any spilling, leaking, pumping, pouring, emptying, discharging, injecting, escaping, leaching, dumping, or disposing any toxic chemical into the environment.

Toxic Release Inventory Database of toxic releases in the United States compiled from SARA Title III, Section 313 reports.

Toxicity The degree to which a substance can harm humans or other organisms.

Toxicology The study and detection of poisons and their effects.

Turbidity measurement Turbidity is defined as an "expression of the optical property that causes light to be scattered and absorbed rather than transmitted in straight lines through the sample" (*Standard Methods for the Examination of Water and Wastewater, APHA, AWWA,* and *WPCF*, 16th ed., 1985). Turbidity is the measure of relative sample clarity.

Urban ecology The study of the relationship of living organisms with each other and with their surroundings in an urban environment.

Urban heat island Urbanized areas that are significantly warmer than the surrounding rural areas resulting in a microclimate.

Urban metabolism The description and analysis of the flows of the materials and energy within cities.

Waste-to-energy facility A facility that incinerates waste and uses the heat energy produced to power generators, creating electric energy that is usable by consumers.

Wellhead protection zone The surface and subsurface area surrounding a well used for drinking water.

Wetlands Areas where the frequent and prolonged presence of water at or near the soil surface drives the natural system—the kind of soils that form, the plants that grow, and the fish and/or wildlife communities that use the habitat. Swamps, marshes, and bogs are well-recognized types of wetlands. However, many important specific wetland types have drier or more variable water systems than those familiar to the general public. Some examples of these are vernal pools (pools that form from the spring rains but are dry at other times of the year), playas (areas at the bottom of undrained desert basins that are sometimes covered with water), and prairie potholes.

Wind farms Wind turbines that are grouped together into a single wind-generating "power plant" to generate bulk electrical power.

X-ray fluorescence (XRF) A nondestructive analytical method that does not require the dissolution and loss of samples or the disposal of hazardous waste solvents, as do traditional wet chemical methods. XRF units can provide a fast, safe analysis of trace elements.

The Metric System

The metric system originated in the 1790s as an alternative to the peculiar, traditional English units of measurement. The metric system has been used internationally in the fields of engineering and science for many years. It was not until the 1970s that there was an international movement for the global adoption of the metric system. Although the United States made an attempt in the mid-1970s to adopt the metric system, the attempt failed, and we remain one of the last holdouts. As students of the environmental sciences, we sometimes find that this dual system causes confusion and problems. Nevertheless, to practice and understand science, you need to know the metric system. And to communicate your findings to the general public, you need to know how to convert measurements into traditional English units of measurement. There are numerous online conversion calculators and most introductory science textbooks have conversions in their appendices. Below are some common metric units and conversions.

LENGTH

1 centimeter = 10 millimeters
1 decimeter = 10 centimeters or 100 millimeters
1 meter = 10 decimeters or 100 centimeters or 1,000 millimeters
1 kilometer = 1,000 meters or 100,000 centimeters or 1,000,000 millimeters

AREA

1 cm^2 = 100 mm^2
1 m^2 = 1,000,000 cm^2
1 hectare = 10,000 m^2
1 km^2 = 100 hectares or 1,000,000 m^2

VOLUME

1 centiliter = 10 mL
1 deciliter = 100 mL
1 liter = 10 dL or 1,000 mL
1 m^3 = 1,000 L

MASS

1 kg = 1,000 g
1 ton = 1,000 kg or 1,000,000 g

Conversion Factors

English to Metric Conversion Table

Change	From	To	Multiply by
LENGTH			
	feet	meters	0.3048
	inches	millimeters	25.4
	inches	centimeters	2.54
	miles	kilometers	1.6093
	yards	meters	0.9144
VOLUME			
	cubic feet	cubic meters	0.0283
	cubic yards	cubic meters	0.7646
	gallons	liters	3.7853
	pints (dry)	liters	0.5506
	pints (liquid)	liters	0.4732
	quarts (dry)	liters	1.1012
	quarts (liquid)	liters	0.9463
WEIGHT			
	ounces	grams	28.3495
	pounds	kilograms	0.453592
	pounds	metric tons	0.000453592
	tons (short)*	metric tons	0.907185
AREA			
	acres	hectares	0.4047
	square feet	square meters	0.0929
	square miles	square kilometers	2.59
	square yards	square meters	0.8361

*Short ton or American ton (2,000 lb).

Miscellaneous Conversion Table

Change	From	To	Multiply by
ENERGY			
	kilowatt-hour	BTU	3,412
	watts	BTU/hour	3.4121
SPEED			
	feet/second	meters/second	0.3048
	miles/hour	kilometers/hour	1.60934
TIME			
	hours	days	0.04167
	days	years	0.00273785
TEMPERATURE			
	degrees F	degrees C	$-32 \div 1.8$

Numerical Prefixes

The International System of Units uses the following prefixes to represent large and small decimal numbers. The following is a list of the more common units you will use in the labs and problem sets:

Positive numbers			Negative numbers		
PREFIX	**SYMBOL**	**NUMBER**	**PREFIX**	**SYMBOL**	**NUMBER**
peta	P	10^{15}	milli	m	10^{-3}
tera	T	10^{12}	micro	μ	10^{-6}
giga	G	10^{9}	nano	n	10^{-9}
mega	M	10^{6}	pico	p	10^{-12}

About the Authors

Travis P. Wagner is a professor in the Department of Environmental Science and Policy at the University of Southern Maine. He received a Ph.D. in environmental and natural resource policy from The George Washington University, an M.P.P. in environmental policy from the University of Maryland–College Park and a B.S. in environmental science from Unity College in Maine.

Robert M. Sanford is a professor in the Department of Environmental Science and Policy at the University of Southern Maine. He received a Ph.D. and an M.S. in environmental science from the State University of New York College of Environmental Science and Forestry and a B.A. in anthropology from the State University of New York College at Potsdam.